REINVENTING
CHINA

REINVENTING
CHINA

The Experience
of Contemporary Chinese
Returnees from the West

Zhuqing Li

Bridge21 Publications, LLC
Los Angeles, California

Reinventing China:
The Experience of Contemporary Chinese Returnees from the West
by Zhuqing Li

©2016 Zhuqing Li

Distributed by Transaction Publishers
10 Corporate Place South, Suite 102
Piscataway, NJ 08854

Copyeditors: Amanda Brown and Jennifer Collins
Layout Design: Linda Ronan
Cover Design: Kiryl Lysenka

For information contact Bridge21 Publications, LLC
16705 Huerta Road, Encino, California 91436 or info@bridge21.us

Published in the United States

ISBN: 978-1- 62643-051- 8 Paperback

Contents

To Ed, Daniel, and William, with love.

The Winds of Change

In the early 1980s, China was beginning to bud with change. This was especially true in the southern city of Guangzhou where I attended Zhongshan University. The moss-covered, dilapidated colonial-era buildings that lined the streets were starting to undergo renovation—much of it slapdash, but a big improvement over the utter neglect so typical of China's built environment at the time. Markets, totally forbidden just years earlier, were now starting to hum, often with a vibrancy of color and style imported from nearby Hong Kong. These colors were slowly but surely starting to show up on college campuses filled with a new generation of young students. Guangzhou was and is the largest southern city in China. Abutting Hong Kong, it has always been a window to the West, a weathervane for China's changing times.

Part of the above-mentioned change involved the reinstatement in 1977 of the national college entrance exam, the famously high-pressure, merit-based system for selecting the precious few who would be given a shot at higher education. The system had been banned as bourgeois and non-egalitarian during the ten years of the Cultural Revolution (1966–1976), and thus a backlog of college aspirants had steadily been accumulating. High school graduates, many of whom had been sent down to the countryside for "reeducation" by the peasants, simply had nowhere to go. But for them at least, the world changed in 1977.[1]

When the gateway to meritocracy was thrown open, a decade's worth of high school graduates—many of whom had probably never thought they'd be given a chance to continue their education—stampeded into exam rooms around the nation. During that first year, a mere 4.5% or so gained access to the limited number of slots in the nation's existing colleges. This was the legendary "Class of 1977," designated by their year of entry. They were the best of the best. And they spanned, in a sense, generations. Uncles and nephews could be found sitting side-by-side in the same classroom.

Similar conditions still prevailed two years later in 1979, the year that marked the official start of Deng Xiaoping's "Reform and Opening" policy that introduced a market economy and privatization, and eventually opened up the country to foreign investment.[2] Nobody knew where it would lead to, but the winds of change had been unleashed. Across the country, college students—both younger and older—were embarking on a new journey of reflection and exploration.

To younger students in the college class, the older ones seemed to have lived in a previous life and realm before they entered the university. Some spoke of reading by their pea-sized oil lamps after long days of physical labor, fighting off cold, hunger, exhaustion, or tropical mosquitos in remote villages; others spoke of their privileged childhoods of promise being replaced by broken families; or of surviving as young adult orphans when their parents and guardians were persecuted to their deaths. Theirs were stories of sudden, incomprehensible, and complete changes of fortune: parents who were esteemed experts one day being turned into counterrevolutionaries the next, and sent away to remote villages to be re-educated; high-school-aged youths who answered Chairman Mao's calls to build the remote countryside, only to find themselves trapped there with no hope of recovering their lives; or naïve children who reported their loved ones' counterrevolutionary misconduct to the party only to find out later that their solidarity with the party had not only wronged their loved ones but also led to their demise. Cruel turns of fate such as these punctuated the coming-of-age stories of many of the older classmates in college, and they would later become reminders of a place and time never to be revisited, as well as a landmark that shows how far the Reform Era has come.

In those early post-Cultural Revolution years, a new literary genre burst onto the scene. Written mostly by newly returned educated youths from the countryside who had witnessed and lived through the darkness and the suffering of that era, it was later called "scar literature."[3] In different literary forms—fiction, poetry, or memoir—scar literature told heart-rending stories that took place during the ten-year period when intellectuals were in exile. If Mao had looked farther ahead, he probably would not have sent intellectuals to the remote corners of the nation, knowing that they would one day return and bear witness to a dark age of despair with lives on the brink of survival and insanity. Among the returned youths who found their voices in the new Reform Era, the Misty Group poets, for example,[4] had perhaps the most distinctive voices conveyed in a new free-style poetic form, from Gu Cheng's contemplation ("The dark night gave me a pair of dark eyes / But I use them to search for brightness"[5])—to Bei Dao's anger, questioning, and sense of historical mission:[6]

> Baseness is the passport for the base
> Loftiness is the epitaph for the lofty
> . . .
> The Ice Age has passed
> But why are there still icicles everywhere?
> . . .
> Let me tell you,
> The world—
> I-do-not-believe!
> . . .
> New turning points and glistening constellations
> Are filling a boundless sky
> They are the pictographs of five thousand years
> They are gazing eyes from the future.

Equally challenging was the emergence of new feminism and the declaration of love found in Shu Ting's "To the Oak Tree":[7]

> If I love you—
> I will not wind upon you like a trumpet creeper
> To dazzle from your height
> . . .
> I must be a giant vermilion flower tree
> Standing alongside you
> With the roots grabbing firm the earth
> And leaves touching the clouds along with yours
> . . .
> My love for you—
> Is not just for your imposing and sturdy stature
> But also for the stand you take
> And the land under your feet.

Both the older, returned youths—such as these writers—and their younger classmates who were fortunate enough to have escaped the ravage of the Cultural Revolution embraced the new world that they had never dreamed of living in.

In the Foreign Language Department where I attended classes in the early 1980s, that new world included the nation's first batch of foreign teachers arriving from the Western worlds of America, Australia, Great Britain, and Germany. They gave lectures on Dickens, Hemingway, and Robert Frost, and offered introductory courses on market economies and the structure of the American government. Around the campus at night there were lectures on subjects such as classical music and Western philosophy, as well as dance parties where many students took their first tentative and excited dance steps.

This was a time of great promise and possibility. Gatherings like these would often end in noodle and porridge shops outside the university gates, where the country's new entrepreneurs—people who would have been vilified, or worse, just a few years earlier—were starting to set up shop. As a new generation of college students, they were the earliest beneficiaries of the nation's new meritocracy. They did not know what the future would be like, but in their youthful exuberance they believed it was in their hands, and they were at the cusp of a great change. Most of all, they dreamed of studying abroad.

But just as this was a time of hope, it was also a time of frustration and despair. Many elements of the old socialist system remained stubbornly in place. Even for those lucky few who made it to college, their future path was effectively controlled by the State. Upon graduation, the government assigned them to particular jobs and particular employers, ostensibly for life.[8] In many ways, despite their newfound freedoms, they were but tools of a state plan. Top college students in fields deemed most important for the country's development were sent for training in developed countries, oftentimes taking opportunities from their equally brilliant counterparts in humanities fields in the same schools. As the Reform and Opening proceeded, many found ways to study abroad with private sponsorship—either from their relatives residing abroad or from foreign institutions—effectively freeing themselves from government control. While studying in the US, many of them—mostly those with private sponsorship but not excluding those sent by the government—found ways to remain in the US.

Those who successfully established themselves in foreign countries found themselves courted by the Chinese government many years later, with salaries and accommodations commensurate with those enjoyed by foreign experts. They had somehow become somewhat foreign. Seeing the Chinese government's eagerness to engage them as a bridge to the developed world, many took these offers as new career opportunities—leaving behind their stable jobs, homes, and families in the United States and elsewhere in the developed world—and returned to China to make new starts.

Upon their return to China, however, many found that while they had pursued their dreams of studying and establishing themselves abroad, some who remained at home had ridden the waves of economic reform and found ways to become millionaires without a college education. Some of these entrepreneurs, perhaps even the noodle shop proprietors of the returnees' college years, were earning even more than the returnees themselves. In this home country that had been greatly transformed and was continuing to metamorphose with great energy and speed, the returnees found they needed to reintegrate, and more importantly, create and define their new roles.

Success and failure, dreams and reality, home and abroad. In the intervening decades, many of these clean juxtapositions have been subject to reinterpretation since the Reform and Opening era transformed the lives of my generation and beyond. In the five stories of returnees that make up this collection, I have attempted an in-depth exploration of their personal struggles and triumphs in forging paths uniquely their own. Why did they return? Why did they stay in China? Have they successfully reintegrated back into Chinese society? Has their Western training played a significant role in this era's transformations? How have they contributed, after their return, to the changes taking place in China? What personal prices have they had to pay along the way, and what have they learned and gained in the process? These are just some of the questions that will help us understand some of the personal stories behind the nation's changes.

To explore questions such as these, I have focused on a number of returnees who have successfully established something new and transformative in the country. Generally, these returnees were college graduates when they left, and were accomplished individuals in America before their return. While most returnees have joined various institutions and become successful upon their return to China, the returnees in the following chapters have emerged as path-blazers. Inspired by their knowledge of the Western world and success in their respective fields, they have held up new visions for their countrymen. This collection consists of stories they have told me as I have come to know them.

A Time-Honored Tradition

Returnees are by no means new to China. The country has a long history of sending people to more advanced places so they can return home with new ideas, models, and motivations. Since time immemorial, influences from abroad have repeatedly transformed Chinese society.[9] Indeed, Buddhism itself, one of the core foundational elements of Chinese religion, came from the outside via a monk, Xuangzang, who in the third century brought the Buddhist sutras from India via the Silk Road. But something that happened much later opened the door to transformation in China that we are still wrestling with today. The nation's devastating loss to the Western powers in the Opium Wars of the mid-nineteenth century jumpstarted a massive and continuous importation of new ideas and technologies, many of which still challenge the Chinese government and society alike. When managed well, these imports—be they abstract ideas or physical technologies—have proven useful to the government's developmental agenda. When misapplied during times of upheaval, they have served as unpredictable accelerants—threatening, in effect, to burn down the house. In virtually all cases, they have proven to be a mixed blessing, often prompting strong reactions on the part of the State. The State either scrambles to improvise in order to accommodate them, or quashes them when they appear threatening. Rarely is the reaction simply passive or indifferent.

In almost all cases, returnees have been the bearers of these influences from abroad. And as bearers, returnees have often exerted their own agency—their own sense of mission and purpose—in setting the terms and conditions under which new ideas may continue to enter. In some cases, returnees have grasped directly for the reins of political power. In 1912, Sun Yat-sen, a returnee aided by followers from overseas, supplanted China's last emperor. In so doing, Sun would come to be associated with the founding of the first modern Chinese nation, the Republic of China. In the ensuing decade, returnees from France and revolutionary Russia who were inspired by Marxism and the promise of transformative social change would found the Chinese Communist Party. And across the first decades of the Republic, writers and literati returned to China with all manner of new ideas, thoroughly upending traditional culture and society.

Sometimes, rather than directly challenging tradition and power, returnees attempted to play a more conciliatory, bridge-building role. Indeed, many intellectuals who left China during the Nationalist rule of the 1930s and '40s returned to Communist China after 1949 and attempted to accommodate themselves to the new regime. They went along with the regime's new rules, dedicated themselves to the Party's developmental vision, and toed the line when called upon to foster its sweeping program of societal transformation. But even in a conciliatory mode, returnees frequently ran afoul of the Party, and more particularly, Mao Zedong and his radical vision. In short order, Maoist political movements came to be directed against intellectuals themselves, and particularly returnees. Many were "struggled against" and accused of being foreign spies, capitalists, revisionists, and counter-revolutionaries. Their integrity was questioned and their Western expertise ridiculed. In many cases, they were physically attacked and suffered grievous bodily harm.

Even with the end of Maoism and the start of Deng Xiaoping's reforms in the late 1970s, suspicions about influences from abroad never wholly disappeared. Nonetheless, the era of Reform and Opening marked an unprecedented period of opportunity for returnees. This was especially true in commerce and education. To attract enterprising students residing overseas, the government set up science parks, offered attractive incentives, and even sent out emissaries to invite established scholars and entrepreneurs to return.[10] Many of the most successful entrepreneurs in China today had their start in these science parks, and enjoy a good working relationship with the government.

Yet, as noted, even in these best of times, conflicts and frictions still inevitably surround returnees and the external influences they bear. Today, a number of China's most prominent political dissidents are returnees. Some, like returnee and Nobel Laureate Liu Xiaobo, are suffering long prison sentences. Others, like returnee and artist Ai Weiwei, live in limbo—sometimes incarcerated, sometimes held under house arrest, and sometimes left to their own devices, but always kept under the watchful eye of the State. Other less prominent dissident returnees also endure the challenge of upholding non-conformist ideas in the face of a profoundly intolerant and unsympathetic State. And the political leadership itself seems torn between aspirations toward modernization and openness and antipathy toward any hint of disorder and social unrest.

Most returnees are not outright dissidents. For better or worse, in contrast to people like Liu Xiaobo or Ai Weiwei, they neither place themselves outside the existing political order, nor challenge it head-on. Nor do they wholly acquiesce. Instead, they engage in continual and deeply nuanced acts of boundary-pushing, negotiation, and experimentation. Sometimes they advance, and sometimes they retreat. But they maintain their focus on the long view, standing by the values they believe in, and pushing them whenever possible through whatever means they can. A good example involves a woman who I cover extensively in this study, Li Yinhe. She is one of China's most prominent and provocative sociologists pioneering gender and sexual studies. For over twenty years since her return to China from the United States, Li has fought for individual rights of self-expression and equality, especially with regard to gender and sexual preference. In the fight for such rights, she puts her faith ultimately in the legal system and the legislature, while at the same time acknowledging the many shortcomings of China's current institutional set-up. In essence, she works within the system, ultimately with the hope of changing that system from within. Li Yinhe has publicly and privately lobbied the government to revise China's marriage laws and criminal code (especially as they relate to sexual practices), while at the same time she has pushed the boundaries by calling for the legalization of same-sex marriage and the decriminalization of prostitution. She is confident more changes are on the way.

Such confidence has grown with China's increasing prosperity and deepening connection to the global economy. Particularly on the commercial front, an increasingly self-assured government is more willing to make use of returnees' help while going forward. Concomitantly, the returnees' confidence in where the country is heading inspires more free enterprise and investment. Moving forward, the government has enacted new policies and laws to pave the way for streamlined international business transactions and cross-border investment projects.[11] For more than a century, the government has been sending students to study abroad, but nothing before compares to what the State is now doing to help relocate returnees and facilitate their establishment in China's business ecosystem. The State extends special accommodations to returnees' families as well, including the provision of special schools for overseas-born children, special housing, spousal employment, and many other perks to facilitate smooth re-assimilation. Similarly on the academic front, the government courts the best Chinese scholars residing abroad, doing all it can to lure them back.[12] To a greater extent than ever before—at least in recent memory— those scholars are now willing to engage in the dance.

At the end of the day, this sometimes collaborative, sometimes tense, and sometimes uncertain relationship generates a sense of hope. Some of my returnee interlocutors use the analogy of a brick wall to describe the change they are trying to bring about. A brick wall can be changed in many different ways. It can be knocked down in one fell swoop and replaced with a new one. Or, it can be taken down one brick at a time. As each brick is replaced, the change to the wall is imperceptible. But in the end, an entirely new wall gets built. As they go about slowly but steadily replacing the bricks, returnees welcome the State's tentative willingness to include those from within their own community in its policy-making process. Access to the State is, of course, good for both ego and business. But more importantly, many returnees feel that, to the extent that they can engage in direct discourse with the State, they can help shape the future of the nation. More than making a profit in the market, many successful returnees quote this aspect as the strongest reason to return, and remain in China. "America has a well-established system," one returnee noted, "so over there you are just a cog in a well-run machine. But China is still finding its way. What I do today here in China has the potential to impact its future course."

Agents of Change

So who are the most influential among China's latest returnees? Demographically, they span almost two generations, yet share a common experience in the final upheavals of Mao's China, and the early promise of Deng's. For the most part, they were college-educated in the late 1970s and early 1980s. They experienced the Cultural Revolution in their childhood or early youth, and then were among the precious few who got a chance to continue on to higher education with the start of the Reform Era. They were ostensibly the vessels through which China would realize the "Four Modernizations" and the nation's persistent dream to catch up with the West.[13] Whether for personal gain or national glory, they unflinchingly took up the mission to build a better, brighter, more fulfilling, and more promising life than their parents and grandparents had endured or even imagined. Many in this generation went off to the West when given the opportunity. Perhaps more surprisingly, many have now returned to China.

Why did they return instead of living "happily ever after" in the West? How, if at all, were they transformed by their experience abroad; what elements of that experience did they take back to China? Upon returning to China, what did they try to achieve? How did they achieve it? How was that experience received by those who never left China, but were themselves also trying to transform the country and improve their own lives? This study explores these questions not through aggregate statistics or broad-brush descriptions, but instead through the lens of detailed accounts of eight individuals who have gone out and returned.

The experiences of these eight individuals, while varied, are bound together by several common threads. Each chose to return to China despite having achieved different degrees of professional and material success in the United States. In essence, they could have comfortably remained abroad. Each of these individuals made the choice to return when doing so was not yet "popular" or "normal." That is, they were pioneering, rather than following, what later would become a trend. And each of these individuals, upon returning, introduced something new into Chinese society—something which would fundamentally alter China's contemporary social landscape. Here are glimpses of these agents of change.

When Wu Zhendong (chapter 3) returned to register his general aviation company in 1993, China hardly had a clearly defined sector of general aviation, and the skies were under the control of the military. Riding the tides of an economy that had been expanding in recent years, Wu became one of the leading voices in pushing the government to carve out more general aviation space. Today, general aviation has been made part of the government's two latest five-year plans. And there is every indication that the government will continue to make more airspace available for commercial and general aviation.

The environmentalist Liao Xiaoyi (chapter 1) returned in 1996 to find China's environmental and ecological system severely threatened by the country's rapid industrialization. Inspired by Rachel Carson and the American environmentalist movement, Liao founded one of China's earliest environmentalist NGOs soon after her return, and named it the Global Village of Beijing. Liao started her work in the nation's capital, mainly basing it on the mainstream American-style environmental protection and conservation measures that promote recycling and energy efficiency.

In the wake of the 2008 Wenchuan earthquake in Sichuan province, she gathered enough resources to rebuild Daping, a village at the epicenter of the quake. Her Beijing-based Global Village is now one of China's foremost NGOs campaigning for environmental protection, and Liao is one of the world's most decorated environmentalists holding some of the most prestigious international and domestic awards for her work.

The newly rebuilt Daping Village is Liao's ongoing new social experiment. She seeks to stop environmental degradation by transforming the community and people's attitudes towards nature. Yet the new eco-friendly houses in Daping are not working as they are supposed to and plans for community building have not always delivered the intended results. And amidst the glowing reports of Liao's achievements, villagers' voices were never heard nor the achievements of her social experiments scrutinized.

Nevertheless, Liao's story is unique because, at a time when much was in dispute between China and the West on the environmental front, Liao made herself a recognizable figure welcome on podiums both at home and abroad and a darling of the media. On the international stage, she calls for China to improve its environmental record, but she also advocates for China's equal rights in pursuing quality of life and the need for modernization. The former is a voice that the developed nations want to hear from a Chinese national, and the latter no doubt endears her to the authorities back home and common people alike. Her story is also unique in that after following the American lead in environmental protection, she was disillusioned with what she viewed as the juxtaposition of man and nature in the Western tradition, and called for the rediscovery of Chinese traditional wisdom in which humans live harmoniously with nature. This switch to traditional Chinese philosophy and the choice of the word "harmony" conveniently coincided with the government's political agenda of seeking social harmony. And she goes one step further to proclaim that the limitations in Rachel Carson's brand of environmentalism can be remedied by traditional Chinese philosophy.

Another story (chapter 4) is about a group of Beijing-based returnees who, in their own quiet way, took on China's system of early education. Childhood learning should be fun, they believed, because it shapes learning habits for a lifetime. Learning shouldn't just be for exams; it should be seen as a lifelong search for enrichment. These particular returnees under discussion—four mothers of young children who in the 1990s gave up highflying jobs as engineers in major American companies—returned with their families to Beijing to start an American-style, non-profit children's library. Theirs was the first such library in the country, named Peekabook. Inspired by their example, many similar libraries sprang up in the city. Now such libraries in the capital city are standing side by side with, and in juxtaposition to, local training centers, so-called "cram schools" that crank out children who have essentially become professionalized exam takers. These libraries offer an antidote through "open stack" access to books of all manner and kind, but also through classes that immerse children (and sometimes their parents) in activities that connect learning with discovery of the world and the self. As an NGO, Peekabook started out by relying entirely on the founders' own resources, profits made in the American hi-tech industry before their return. Unlike many who returned to establish start-ups hoping to profit from the Chinese economic machine, these founders are making a cultural statement.

This story is as much about the Peekabook library as the founders themselves. All in their mid-forties, they face the same challenges as other professional women: finding a balance between professional pursuits and the responsibilities of raising a family. This challenge took on an additional dimension when they gave up their successful jobs in the States and returned to start anew in China. Without promotions or salaries, the only validation of their work has been the social impact of their library. Having found smooth sailing until midlife, these women face these challenges together, building strong bonds in the process. With this library, they want to showcase the advantage of the American style of education as an alternative to China's exam-oriented approach. Raising their own children in this context has become their own experiment. In their collective search for a balance between this work and raising a family, they are experimenting with a combination of what they see as the best of both cultures. Today, the learning style they promote with their library has grown in popularity. With an increasingly open society and the deepening of globalization, this social experiment is becoming a viable alternative style of education.

Women, self, and self-identity are some of the central topics that sociologist Li Yinhe (chapter 2) has been exploring since her return to China in 1988. She is often called the first and only female sexologist in Chinese history. For Li Yinhe, returning to China after getting her sociology PhD from the University of Pittsburgh was simply a professional necessity: "I'm a sociologist, and China is my society." The society to which she returned was taking off economically, yet still with entrenched traditions and customs. Amidst the onslaught of drastically different Western values, lifestyles, and other cultural imports that have arrived with advanced Western technologies, the government has no directives on how to reconcile them with age-old Chinese traditions. Li Yinhe moves into this vacuum and presides over it. She is convinced that only a more open society can accommodate a higher degree of industrialization and integration into a global economy. In her view, the key to this open society is the awakening and assertion of self. As she tells her fellow citizens, it is a human being's fundamental right to assert oneself and make one's own decisions. This includes expressing and handling one's own feelings, sexual desires, and preferences. In a society where group interest has always trumped personal identity, and where these personal issues have always been taboo in social discourse, Li's advocacy upends the tradition. To conservatives, Li has desecrated the age-old tradition of decency and has replaced it with the decadence of Western society.

A prolific and outspoken scholar, she stands her ground. Using modern social media, she fights to validate the most unspeakable personal desires and fears, lending a voice to and providing a forum for those who never knew how to express such matters. Furthermore, she takes action to effect political and legal changes that safeguard progress made. Though many battles remain to be fought, profound change has occurred. In my own college years in the early 1980s, campus police scoured the shrubbery on campus to rout out young couples doing nothing more than trying to steal a kiss. Today, Chinese society is publicly debating the legalization of gay marriage and the decriminalization of prostitution, and the government consults with Li Yinhe on revisions of marriage law. In an interesting way, the lines have blurred between her personal achievements and the nation's progress.

The blurred line between the individual and the nation's narratives is deliberately erased in the next story. Here start-up guru Chen Datong (chapter 5) puts his own path to success alongside the Chinese Communist Party's road to power, and declares: "We are not witnessing history, we are making it." Chen, who has founded successful companies in both China and America, is one of the key people behind the development of the inexpensive smart phones that transform life for ordinary citizens within and beyond China's borders.

Chen returned to China to found Spreadtrum after his first company in Silicon Valley, OmniVision, went public. If

OmniVision is a typical start-up success story in America, Spreadtrum may be a textbook case of "disruptive innovation" in China. Coined by the Harvard professor Clayton Christensen in his 1997 book, *The Innovator's Dilemma,* disruptive innovations compete with established products by means of lower profit margins and lower quality products that capture lower market tiers. There is a folksy name in Chinese for such a disruptive model: *shanzhai,* meaning "mountain stronghold." This rebellious nickname took off along with the rapid spread of the *shanzhai* cellphones, and has now become a modifier for many other products and even events. Some may still debate the difference between disruptive technologies and plagiarism. The Christensen Institute is unambiguous in its definition of such technology: "It's important to remember that disruption is a positive force. Disruptive innovations are not breakthrough technologies that make good products better; rather they are innovations that make products and services more accessible and affordable, thereby making them available to a much larger population."[14]

Yet when Chen talks about how he and his classmates in graduate school successfully reverse-engineered a Texas Instruments chip, one wonders whether it was just a learning process or a practice that infringed on intellectual property rights. As he is careful to point out, they did not use this chip to make a profit. It is indeed increasingly difficult to distinguish between learning processes, plagiarism, disruptive innovation, and nationalistic pride in this reverse engineering experiment. Lines between these notions in the country's breakneck drive for economic power and technological superiority seem to be drawn in the sand. But Chen is not bashful about the rebelliousness of either the product or its "mountain stronghold" name. Moreover, he pushes this rebelliousness to a new height. In Spreadtrum's success in every step of their progress—being the first to adopt China's own standard, bringing the country's poor into the telecommunications age, and challenging the domination of multinational giants—Chen sees a parallel to the Chinese Communist Party's ascension to power in the 1930s and '40s. In Chen's recounting of Spreadtrum's success, one feels that it is not only about how a company succeeded in China's telecommunications era; it is more about one man's quest to establish a legacy and find meaning in the context of a greater cause. In his attempt to do so, he frames his company's success with the narrative of the CCP's victory and its ultimate establishment of a new country.

Indeed, history has been made. People in different camps with differing ideologies and cultural and political views are scrambling to put a name to every new thing and every new practice and phenomenon as they keep popping up. In China, the millennials—the generation born between 1980 and 2000—are living in a transformed world. State control over every aspect of an individual's life has given way to considerable freedom and choice, with the caveat, of course, that political expression is still largely off limits. College graduates are now free to find—indeed, in order to survive, they absolutely must find—whatever jobs they can. They choose where to live, what health insurance to buy, and what school to send their children to. The choices are many, and guarantees few and far between. But to the enterprising, the opportunities seem limitless.

For these millennials, the world outside China is no longer glimpsed through library books or occasional lectures and interaction with foreign teachers. At least for educated city dwellers, this generation is comfortable with a foreign language (usually English) and comfortable in a global culture of online interaction, social media, and ubiquitous smart phone use. Children from more affluent families now go abroad for vacations, summer camps, and private secondary schools. Unlike my generation that had to rely on outside sources for tuition and living expenses while studying in the United States, today's prospective students from China have become courted targets of many American schools, in no small part because many of those students come from families that are able to pay full tuition.

For those who left China in the 1980s and early 1990s, the dawn of the Reform Era, there was no way they could have imagined what the world of the present would look like. But without a doubt, they dreamed of a better society modeled roughly along the lines of the advanced nations of the Western world. Today,

China is approaching, and even rivaling, a number of those nations on the economic front. And Chinese society, much like the still more advanced nations of the West, is now suffering the ills of wealth: diabetes, obesity, environmental degradation, and income disparity, just to name a few. Debates understandably and appropriately persist over China's ongoing authoritarianism and suppression of human rights. But in many respects—socially, economically, and even to some degree politically—the playing field between East and West has been leveled to a greater degree than ever before, to both good and bad effect. At the very least, for the first time in Chinese history, the nation has truly globalized.

Successes of the present pose as much of a challenge to today's educated youth as they did to my generation. The "one child family" generation of the millennials has aspirations for the future, just as we did.[15] As I compiled data for this study during my year-long period of residence in Beijing in 2013, I met many citizens. Some were privileged, others poor, some educated, and others not. All of them seemed to have an opinion about China's future, and many project their futures in the context of the greater world. For example, a young security guard struck up a conversation with me in English as we ran laps on the Tsinghua University track. He told me that his dream was to get a law degree in America, because he believed revising the constitution was crucial for China's future. And a utility worker who came to fix the pipes in my apartment was intensely curious about all things in American life. Point by point, he analyzed for me which pieces of Americana would be good for China, and which should be avoided. Many more people with varying degrees of education whom I have met talked about different ways to improve the country and are taking ownership of the nation's future.

China has come a long way since Chairman Mao called for the Chinese people to be their own masters. After answering Chairman Mao's calls to "overtake" England and "catch up" to the US (超英赶美) in the '50s and '60s, people now feel confident enough in this Reform Era to claim the country's future as their own. The challenge today is that there is no clear superpower to overtake, and no society so much more advanced than China that it is obviously worth emulating. The challenge for those who search now is to define a clear goal. Well-trained and familiar with Chinese and foreign cultures, the returnees have taken the lead to try out something new on a grand, nationwide scale. I have chosen to focus on the returnee experience primarily because I feel it is exerting an outsized influence on changes unfolding in China today. Without question, China's media have closely followed their social experiments as high-profile examples of reform. In many ways, perhaps part and parcel of the elitism and reverence for educational achievement that has long pervaded Chinese society, returnees are held up within the nation as a group worthy of attention and emulation. For these returnees themselves, after they have introduced and established something new in their native country, they have taken it upon themselves to define their achievements in the context of China's modernization. In every one of the cases discussed in this work, accompanying their confidence is a sense of historical mission and the practice of viewing their accomplishments in China's larger historical context.

Among the returnees who have met with the most success, one can discern certain shared qualities. First, these individuals were well-educated and solidly rooted in Chinese culture before they left the country for America. This rootedness facilitated their reintegration into Chinese society upon their return. They were also able to identify from their American experience what China needs. But, more importantly, upon their return to China, they all seem to have come to terms with the realization that superior training and advanced degrees are not enough to introduce something new and make it stick. Successful returnees knew they would need, at the very least, the government's tolerance, and perhaps even its overt support. They had to negotiate some kind of relationship with the State. More often than not, they avoided the usual approach of wining and dining, flattery and blandishment. Instead, as highly regarded experts in their respective fields, and as people whose skills would allow them to renew careers outside of China if they so wished, they were able to negotiate on terms that more nearly approached equality. The dance between people and government has never been easy, and given the power of the State, it's never entirely fair or equal. But neither is it one simply of subordination, acquiescence, and obedience.

As China continues its integration into the world, these returnees continue to be the bridge between their native country and the world that exists outside it. At home, they are spearheading changes in their fields, as well as influencing government policies; abroad, they are often the faces and voices representing China. This is not the first group of returnees to affect China's development. But, to a greater degree than ever before, the nation is giving them an open platform. They are stepping up to run the show as they claim a piece of history.

Introduction Notes

1. On October 12, 1977, the State Department issued "Opinions on Admission to Higher Education in 1977" 于1977 年高等学校招生工作的意见. This set the ground for the reinstatement of the college entrance examination on merit instead of by recommendation, a practice enacted during the Cultural Revolution. In the winter of 1977, 5,700,000 prospective students applied and 278,000 were admitted; and out of the 6,100,000 applicants in the summer of 1978, 402,000 were admitted. In 1977–78, a total of 680,000 new students were admitted to the nation's colleges. See Long, Pingping 龙平平 and Zhang, Shu 曙, *News of the Communist Party of China*. http://cpc.people.com.cn/GB/64162/64172/85037/85039/6032327.html (accessed March 29, 2016).

2. On December 18–22, 1978, the Third Plenary Session of the Eleventh Central Committee of the Chinese Communist Party 十一届三中全会 set the country on the course of Reform and Opening, marking the beginning of this historic era. See Central Government Information Repository 中央改革信息. http://www.reformdata.org/content/20121106/1438.html (accessed March 29, 2016).

3. Scar Literature: "the term refers to works in the late 1970s and early 1980s that portray the psychological 'scars' that resulted from the Cultural Revolution" (Sun Chang, Kang-I and Stephen Owen, eds. *The Cambridge History of Chinese Literature*, vol. II, *From 1375*. Cambridge University Press, 2010, 651).

4. The Misty Group 朦派 burst onto the scene in the late 1970s and early 1980s. Represented by leading poets—including Bei Dao 北, Shu Ting 舒婷, Gu Cheng 城, Jiang He 江河, and Yang Lian—the Misty Group acquired its name with its new poetic form and unusual accompanying imagery

and structure. A good introduction to this group and the poet Bei Dao can be found in the introduction to *Notes from the City of the Sun: Poems by Bei Dao*, ed. and trans. by Bonnie S. McDougall (Ithaca, New York: Cornell University East Asia Papers: China–Japan Program, No. 34, 1983).

5. Gu Cheng 城 (1956–1993) is one of the leading poets of the Misty Group. He was born in Beijing and was the son of a well-known poet, Gu Gong 工. Years after he emigrated to Auckland, New Zealand in 1993, he committed suicide. The lines quoted are his best known, from the 1980 poem "One Generation 一代人: 黑暗了我黑色的眼睛，我却用它找光明," here translated by the author.

6. Bei Dao 北, ne Zhao Zhenkai 振 (1949), is one of the leading Misty Group poets. He was chiefly responsible for setting up the magazine *Today* 今天 that launched the Misty Group onto the national scene in the late 1970s and early 1980s. The first lines quoted here are his best known, from the poem titled "Answer 回答 (1979): 卑鄙是卑鄙者的通行，高尚是高尚者的墓." For more details on Bei Dao, see note 4 above.

7. Shu Ting 舒婷, née Gong Peiyu 佩瑜 (1952–), is another one of the leading Misty Group poets. "To the Oak Tree 致橡" is her best known poem. Written in 1977, this is generally believed to be the first love poem published after the Cultural Revolution.

8. Prior to 1985, all college graduates were assigned jobs by the government. This changed when, in May 1985, the government issued the CCPCC (Chinese Communist Party Central Committee)'s "Decision on the Reform of Educational System 中共中央于教育体制改革的决定." The Decision proposed an eight-point reform, including ending the system of job assignment for college graduates and giving universities more freedom in management. The complete text can be found on the website of the Ministry of Education: http://www.moe.edu.cn/publicfiles/htmlfiles/business/moe/moe_177/200407/2482.html.

9. For the latest comprehensive survey of the history of Chinese citizens studying abroad, please see Wang, Huiyao 王耀, *Blue Book of Global Talent: A Report on the Development of Chinese Studying Abroad (2012) No. 1*. 国际人才皮：中国留学发展报告 (Social Science Archive Press 社会科学文献出版社, 2012).

10. In 1990, the National Conference of Entrepreneurship 全国业中心第一次工作会 was held in Beijing; in 1997, the first entrepreneurial park for returnees (留学人业园) was set up in Haidian district in Beijing. By 2007, there were 23 of these parks for returnees, attracting 3,372 returnees and more than 2,000 startups. Between 2000 and 2007, these parks had a total income of 16.1 billion yuan, paying taxes of close to 700 million yuan. (*People's Daily* (Overseas edition) January 10, 2009, http://paper.people.com.cn/rmrbhwb/html/2009-01/10/content 173333.htm 中国新网 http://www.chinahightech.com/html/684/2014/1020/936363336353.html).

The Office of Educational Affairs of the Embassy of the P.R. China in the US offers a set of comprehensive guidelines regarding the country's support and accommodations for those who choose to return. It includes sections that list the major programs set up to attract the best scholars and the available funding, etc., complete with numbers and statistics:http://www.sino-education.org/policy/studybrief.htm.

11. For information on commercial law reform, see Dong, Yuming 董玉明. *In Step with Reform: Theories of Commercial Law and the Study of Time Management* 与改革同行：经济法理与时间研究 (Intellectual Rights Press 知出版社, 2007.), Also see the textbook published by Qinghua University, *On Commercial Law* 经济法概. 2004.

12. For details of the Chinese government's regulations, see the website of the Office of Educational Affairs of the Embassy of the P.R. China in USA: http://www.sino-education.org/policy/studybrief.htm.

13. The term "Four Modernizations" is generally believed to have been mentioned by the former premier, Zhou Enlai, on January 29, 1963. Premier Zhou Enlai proposed at the Shanghai Science and Technology Meeting that China "should realize agricultural modernization, industrial modernization, modernization of the national defense, and of science and technology." This was later summarized as the Four Modernizations (News of the Communist Party of China, http://cpc.people.com.cn/GB/64162/64165/76621/76651/5289691.html) Deng Xiaoping in his various talks and speeches renewed the call for the Four Modernizations at the beginning of the Reform and Opening era.

14. See the Clayton Christensen Institute website at http://e360.yale.edu/mobile/feature.msp?id=2782.

15. The "one child families" are the result of China's family planning policy that started in 1980. In the 1980 government document "An Open Letter to All Members of the Communist Party and Communist Youth League on Controlling Population Increase 于控制我国人口增长致全体共党共青团的公信," the government proposed that each couple should have one child only. The following year, the Family Planning Committee 国家划生育委会 was set up. For further reference please see: Liang, Zhongtang 梁中堂. "Arduous Journey: From One Child to Homes of Girls" 的历程：从 "一胎化" 到到女儿户." *Open Times* 放时代 (online magazine). http://www.opentimes.cn/bencandy.php?fid=375&aid=1806.

Another Lifestyle is Possible:
Liao Xiaoyi, Environmentalist

Daping Village

After the Great Sichuan earthquake of 2008, many learned that the epicenter was home to giant pandas and that thousands of school children had perished in the collapsed school buildings.[1] None in the outside world, however, learned of Daping, a small village located in the hardest hit area. When the earthquake struck, almost all the able-bodied were in the mountain tending plots of *huanglian*, a plant whose extremely bitter root is prized in Chinese medicine. Over ninety percent of the village's houses collapsed, yet their livelihood survived, as did ninety-five percent of its population of over 200 people.[2] Two years after the devastation, all the villagers were settled in brand new, traditionally styled, eco-friendly homes, mostly connected to the village's first-ever paved road. This was all made possible by a middle-aged woman named Liao Xiaoyi ("Sheri"), one of China's leading environmentalists and social entrepreneurs. Like all others in this collection, she is a returnee, a Chinese citizen who studied in the United States and returned to China with new ideas and approaches.[3]

Liao saw the earthquake as a chance for Daping villagers to rebuild a better life. "Bitter like *huanglian*" is the old Chinese saying. It may be as true of the root's taste as the backbreaking

31

work of cultivating it. Villagers in Daping eke out their livings on very steep, cool, and perpetually misty slopes at an elevation of 1,600 meters. For each of the five years that *huanglian* takes to mature and be prepared for market, a family gets an annual income of over 10,000 yuan (about 2,000 US dollars). Liao thinks they could do better not just financially but also in terms of their quality of life.

On a rainy, cold evening in the late autumn of 2012, Liao proposed a new plan to generate more revenue for the village to Mr. Yuan, the Party secretary of Daping. The plan included the following: a health spa; the life-enhancing workout devised by Liao's Chinese medicine practitioner; organic food produced right in the village, the production of which would be directed by her organic farming advisor; and arrangements for handicrafts and other entertainment for tourists' leisure activity. She wanted Secretary Yuan to pick some people from the village for her to train for such an enterprise.

Secretary Yuan was slow to take up the new idea, telling Liao: "You've got us to try many of those things in the past four years with little success." Moreover, he didn't like the large turnover rate in Liao's team of six or seven young volunteers. "They need to stay long enough to finish the things they started," he said. Yet in the end, Secretary Yuan leaned toward giving her the benefit of the doubt one more time. He explained his reasoning to me afterwards: "She has done good things for us, and she means to do good."

The Global Village in Beijing (GVB)

When Liao founded the Global Village in Beijing (GVB) in 1996, it was one of the earliest registered environmental NGOs in China. Its originally small office has now been replaced by two mirror-image apartments in the city that have been donated to Liao's cause. The unadorned entry still has the two opposing doors of the original apartments, which force a visitor to choose which one to enter. The first time I visited, I chose the one on the right, and I found myself confronted by three unfamiliar photos in the entryway: images of Rachel Carson, Liang

Shuming, and Suolandajie.[4] This side happens to be the part Liao uses most often. The bedrooms to the side are her living quarters, and the large living room serves as a meeting space. Had I chosen the door on the left, I would have been greeted by one of the volunteers who works at the front desk in the office. Four or five more work in cubicles that form two clusters in the living room where campaign T-shirts, flyers, and shelves packed with their publications covered the wall space.

On my first visit to Liao on a chilly morning in late fall, the sparse meeting room had the feel of a Confucian temple. A life-sized wooden sculpture of Confucius stood against the wall and faced the entrance: long-bearded and leaning on his stick with one hand, his other hand raised with index finger in the air as if making a point. Behind him, an ink-painting scroll of Confucius in a similar posture appeared like his own shadow. Across the central seating area was a scroll of large calligraphy: "Heaven and Man are One." Underneath this stood a traditional seven-string instrument favored by scholars of old called a *qin*.

The only quiet place that day was Liao's corner bedroom, as a nearby neighbor's dwelling was undergoing renovation. The morning sun was shining on her nightstand when we entered and highlighted a very pretty face in a picture frame. Liao said lovingly, "That's my daughter. Whenever her friends tell her that she is pretty, she'd say that she looks just like her mom." Indeed, that round face bore a close resemblance to the Liao depicted in her earliest documentaries. Now the creases on the mother's face have eclipsed the radiance of youth, turning once innocent bright eyes into a penetrating gaze with thin lips held tight in a straight line either holding back thoughts or ready to unleash a smile. Her voice is gentle and calm, but rises a notch to make a point.

We sit down at a classical scholar's long desk near the bed. A tea set sits on top, next to a lamp. Environmental protection is still relatively new in China, so I start by asking Liao how she got into it. "Ah, let's not talk about the past. That gets tiring." Liao seems a bit impatient, saying, "Those are some 'old sesame and rotten grains'!" Her voice rises somewhat to dismiss the suggestion. She wants to start with 2008. "No, no, no," I insist, couldn't she at least give me some of the stepping stones that led to that point? She looks at me ponderously for a moment, and then relents. But it will be brief.

She begins by saying her first exposure to life in the countryside starting in 1971 was important to her. She was an intellectual youth at age 17 as the Cultural Revolution took hold. It was a time when her "frame of mind was completely broken down by the new reality." She added she now calls herself a new intellectual youth who, having lived in Daping Village for two years, has "found the tree of life."

She explains that her academic background was important. She received an undergraduate degree from Sichuan University and a master's degree from Zhongshan University. Later, she returned to teach in her hometown at Sichuan University and worked as a researcher in the Academy of Social Sciences, all in the field of philosophy. That training and those experiences helped her think and produce over 100 documentaries, numerous articles, and talks that run the gamut of broadcast media. And after 2008, she put her theory into practice in the countryside.

"Wait," I say, pulling her back. "Let's not get too far ahead just yet."

"Right," she says. "And then, America."

In 1993, she went to America to join her then-husband who was studying there for a degree. She took along their three-year-old daughter, Karla. While studying international politics at North Carolina State University and working part-time, she could not forget startling facts she learned shortly before leaving concerning the devastating impacts of China's industrializing consumer economy. Her faith in a market economy was fundamentally shaken.

In America, Liao found Rachel Carson's *Silent Spring*.

Instead of finishing her degree, Liao made her first film, *Daughter of the Earth*. The film debuted in the NGO forum at the 1995 World Women's Conference in Beijing, winning Liao $40,000 from the Ford Foundation. The following year, she gave up her permanent residency in the United States and returned to Beijing to register the Global Village in Beijing. Gliding over it lightly, Liao says the first stage of her environmental work was pretty much along the usual lines of the Western environmental movement.

In the mid-1990s, China was still on the verge of its spectacular economic takeoff, and the environmental impact of rapid industrialization in China had yet to be seen and felt by citizens. Liao, then in her early 40s, started a special TV program called *Environmental Protection Time* on the state-sponsored CCTV, which aired every Friday afternoon. It lasted five years, and was twice awarded the government's special program award. She engaged policy makers and helped draft China's first ever government-issued "Guidelines for Citizen's Environmental Responsibilities," and she worked on garbage recycling, energy saving, and carbon reduction efforts while continuing her media campaign. Her young daughter, Karla, had been constantly cared for by friends around the city. Over a decade later, Liao was still moved when she recalled how Karla saved her mother's favorite food for her till the end of the day when they would meet up.

What had prepared her for the 2008 earthquake was a search that grew out of her deepest despair at the height of her rapid ascendance as a noted environmentalist. She won the Sophie Award from Norway in 2000, and the Banksia Award from Australia in the following year. These high-profile awards put her in the spotlight on the international stage.[5] Instead of enjoying these accolades, she confronted a daunting fact: The world wanted more environmental work done in China, yet there were few people in the field to compete with her for these prizes. The road was long, the task formidable, and the results slow to bear fruit.

As the country moved resolutely toward an American-style society, Liao saw that "Environmental protection itself had come to a dead end, because the modern lifestyle was built on human desire. The 'teacher' didn't help. Carson herself did not solve the problem." Her despair was so complete she even thought about taking her own life to reduce her impact on the Earth, and she told her own daughter not to have children for the same reason.

From 2002 to 2004, Liao took her daughter out of school and the two went on a world tour to seek advice from environmentally conscious scholars. These interviews were made into a book: *Turning to the East and Looking to the West—Environmental Remedies: Sheri Liao's Talks with Eastern and Western Thinkers*.[6] In these interviews, she "sensed that many of these famous scholars also shared [her] frustration that the currently available environmental protection measures were just 'scratching the itch from outside the boots'," and the project helped her to learn the thoughts of leaders from very different backgrounds and perspectives.

Handing me a copy of the book to keep, Liao sits down at the *qin* by her bed. Stroking the strings and softly singing a tune, her thoughts became more upbeat. "With my philosophy background," she says, "I like to think." At the end of her search she had a revelation: "It's the heart of people, the greed, the unlimited desire to extract from nature. It's the juxtaposition of man and nature and treating nature as an inanimate object that leads man to destroy it while satisfying greed. So this is the root of the problem. The West can't solve it, but our national learning, *guoxue*, our own cultural inheritance, can offer answers. Taking the holistic view of traditional Chinese philosophy, man is but a part of nature. If we can fix the heart of the people, we can fix the environment. . . . China does not have to follow Western industrialization to its decline. Another life is possible. We all need to find the shoes that fit us, and put down steady footprints."

Our meeting time is up. But instead of rushing out to the conference she will attend that afternoon, she says, "Time to do *yangshengcao!*" Liao then walks over to the office and rouses the staff from their desks to do their first set of life-enhancing exercises. Her day is structured around performing these exercises four times. As she explains, there are six movements each time that add up to a total of twenty-four *jieqi* in the traditional Chinese seasonal division.[7] Although the volunteers are all around her daughter's age, they cannot match the elegant and precise postures that Liao's slight and often weary frame snaps into. The exercises help everyone stay warm, for Liao delays turning on the heat in the apartment till as late in the season as possible.

When the exercises are over, she changes out of her rustic

cotton outfit and into a well-fitting traditional silk blouse and matching pants. She slips on her pumps and rushes to catch a cab to the conference.

Lehe: Liao's Social Experiment

The 2008 earthquake was still very much on Liao's mind the second time we sat down for a conversation in her home office. When the 8.0 magnitude earthquake hit Sichuan, her home province, she said she rushed there to help in several villages as an NGO leader. In the village of Daping, she saw a few traditional houses still standing among "all the trashy new buildings that had been turned into rubble." In those few traditional houses, Liao saw a symbol of a neglected tradition where man co-existed with nature. She immediately brought in volunteer architects to replicate these homes and equipped them with energy-saving stoves. And she brought in 3.8 million yuan (the equivalent of 600,000 US dollars) from the Red Cross and the Nandu Foundation to build a house for each of the over 90 families in Daping Village. For two years, while living in a tent like everyone else, she ate and worked with the villagers to rebuild as "the community came together, building each other's homes just like their own." Liao still gets emotional recalling those days. In such a society, she said, she believes no one would have the heart to desecrate nature, their common home.

It is this heart Liao has been trying to bring out in people through her social experiment. Liao credits Al Gore for framing our environmental problem as a moral issue. But while Gore stops there, Liao finds guidance in Chinese philosophy. She points to two concepts in particular: "Being in sync with nature, heaven and man in one," and obeying the call to "respect nature and treasure all things." In addition, she finds a model for practice in Liang Shuming's work.

Liang was one of the leaders in the "rural reconstruction movement" of the 1930s. The movement worked in areas of political and economic reform, education, medical improvement, and self-defense. It reinvigorated traditions. Although their operation—some 600 organizations located in over 1,000 experimental sites—was cut short by the outbreak of the war with Japan in 1937, it was credited with laying the foundations for eradicating illiteracy and polio while furthering economic development.[8]

In her own practice, Liao mapped out five fronts: *lehe* dwelling, *lehe* livelihood, *lehe* rituals, *lehe* environmental treatment, and *lehe* life enhancement. The word *lehe* means "Delight in Harmony." When Daping villagers moved into their new homes two years after the earthquake, Liao expanded her Lehe Home philosophy to Yangqiao Village in Wuxi County, also in Sichuan province. "China's future is in the countryside," Liao declared. Despite the urbanization taking place all around the country, she added, China's agrarian history was fundamentally different from the Western commercial tradition. She said that she was sure China's own tradition and practice would ultimately carry the country through its environmental crisis.

The year 2008 was notable for more than just the earthquake, for this was also an important year for Liao personally. Earlier in the year, she had been appointed as one of the two environmental consultants for the Beijing Olympic Organizing Committee. In September, she left the tents in Daping to fly to New York to receive the Clinton Global Citizen Award. At the ceremony, she reiterated that a life of harmony (*lehe*) was possible, and promised she would take the prize as a symbol of affirmation and encouragement for her work back to the village as an inspiration for all. Two years later, Daping Village was reborn.

Buoyed by Daping's quick building reconstruction and the international attention and recognition she received, Liao ebulliently moved on to promoting a higher-standard, self-sustainable, and eco-friendly model for Lehe Livelihood. One specific goal was to arrest and possibly reverse the population outflow into the cities and towns by setting up profitable enterprises in the village as a first step toward building a stronger community. As a journalist reported in October 2010, Liao told one young man in the village: "Once the enterprises get going, you won't have to leave the village to work. You'll just stay home and do the book-keeping." Yang Guanglin, the head of Daping's Eco-association, was equally enthusiastic: "In the future, Daping Village will have a new model: eco-planting and eco-livestock growing. With the latter fertilizing the soil, we could get organic planting going and sell the product to the cities for good prices."[9]

But this dream remains a dream today. Another effort to generate revenue, the "green handkerchief" project, did not last either. Because Liao paid the village women 20% over the market price for each handkerchief they embroidered, she eventually had to abandon it.[10]

Still, Liao was in Daping to stay. She had committed to at least 15 years of partnership with the village to put her Lehe Home philosophy into practice. To her, the social experiment is in itself a significant enough project. As she says, "If we succeed in doing this, we'll make history." She joined the management of the village enterprises, lending them her name for marketing purposes. But the advanced technology involved in the latest project of organic pig-raising makes it an exclusively Global Village (GV) volunteers' job as the villagers do not have enough education to conduct the research and implement the technology involved in the process. To eventually turn it into a village-run enterprise, it will be necessary to train the villagers.

In addition, she decided to expand her social experiment to another county in Sichuan, Wuxi. The building of Lehe Home in Wuxi is associated with a more established bureaucracy. The county has a population of 500,000 people living in 30 towns and villages. The location of their first project, Yangqiao Village, has a population of 5,000. The first year's projected plan included using more than 1,500 mu (20 acres) for organic rice farming plus projects for cleaning the environment, building a dam, greening many areas, and setting up tourist reception sites. The local community responded with enthusiasm, collecting donations of more than 2,600 yuan (almost 300 US dollars) and erecting a Lehe gate at the entrance to the village, using only local materials and volunteer work.

"Let's Turn Our Attention to the Countryside!"
Over the years, many parts of Liao's inspiring projects have brought accolades in the media, as Liao actively promotes her

style of environmental protection by rejuvenating rural communities through reuniting migrant families, rebuilding their homes and communities with sustainable and profitable enterprises, and restoring the countryside into a pristine base for organic products, thus serving as a standard bearer of traditional values. All these efforts intend to reverse the trend of urbanization that is sweeping through the entire country. That one woman has been able to set in motion such a bold vision for the future inspires me to understand how she does it. I believe the heart of her story lies in her work with the Lehe villages in Sichuan.

Both Daping and Wuxi are still in the early stages of this social experiment, and Liao spends much of her year shuttling between these two places. In her latest trip back to Sichuan from GVB in the late fall of 2012, Liao agreed to let me come along. En route, we made a brief stop in Chengdu, the capital city of Sichuan province. As a native daughter turned celebrity, she had been invited back by the Chengdu Public Interest Foundation to present the first awards for the country's microblogging competition for NGOs.[11]

The most important platform for her on that day was a small luncheon for about ten people held at a restaurant overlooking the event site. Local public intellectuals at the luncheon included Fu Yan, the head of the hosting organization, and her young assistant; Wang Keqin, another environmentalist who was also *Chengdu Daily's* editor; and a well-noted pundit named Zeng Ying. Liao first put Fu Yan to task and she agreed to use her microblogging network to help Liao sell the organic pigs and chickens that would be ready for the Spring Festival in a few months. "Let's turn our attention to the countryside!" Liao exhorted. She recalled how she had despaired of doing environmental protection in the "cement forest" of a city where people do not even know their next-door neighbor enough to care for the little stair landing they share. How then, Liao asked, could they care for the Earth? And she made a distinction between acquaintance-based public interest work in Chinese society and the professional public interest work in the West. The latter, she pointed out, offered service, but lacked heart. *Guanxi* (interpersonal relationship) is what Chinese society runs on. While she lamented that "this culture has now been corrupted," she also stated that she believed much of it could still be found in villages. Zeng Ying concurred: "Our education is a complete failure in this respect. It teaches kids that the countryside is a place with no culture or future, a place to run away from."

"Like any experiment, it needs initial investment," said Liao, referring to her community-building in Daping. With funding only for the initial two years, and with the project well into its fourth year, Liao said, "You don't understand how hard it has been. There were times when I could not go on . . ." her voice caught for a moment. "But Daping cannot die. It's a model that has offered us hope. . . . Let me ask you, how many of today's hundreds of NGOs have helped make a positive change in so-cial structure? None, but we do." At this point, Fu Yan's young assistant commented that even the barking dogs in Daping did not bite, presumably making the point that even the dogs there were friendly.

In the afternoon, Liao turns her keynote speech into a rally. She shows a short documentary to introduce herself, shares with the scores of young volunteers some of the ditties she wrote, and brings them to their feet while doing a set of life-enhancing ex-ercises right there in the hall. "Western environmental protection work does not talk about life. But that's exactly what it is all about," Liao told the young volunteers. She explains that West-ern physicists had discovered that only 5% of our world consists of detectable objects; another 25% is called "dark matter," which she equates with Chinese medicine's *qi*, the energy of life; and the remaining 70% is "dark energy," a force she believes is the power of the heart and mind. "If you prostrate in front of God," she says, "you might feel abandoned. But put God inside you and you'll feel different. If this God is the public interest work that you do, then it'll always infuse your life with energy and meaning." This is how she gets through rough patches, she tells her audience. "I'm here to find a resonance today," she says. And she does. When the lyrics of a popular song, "Firefly," shows up on the screen, the hall breaks out in song, and Liao does an impromptu free-dance.[12] "Lives are meant to be set on fire," she calls out to the youth as she ends the rally.

Before getting to Daping, Liao makes one more stop to call on the principal and founder of a local technical college. She brings along her Chinese medicine practitioner and her organic agriculture advisor, two friends and supporters who have offered free advice and tutoring for her team. The purpose of the meeting is to set up a training program in the college as a platform to explore village-building.[13]

The principal, Yan Hong, is a well-known visionary and entrepreneur. She asks Liao a pointed question: Is her vision what the villagers want? Projects such as having rice producers do embroidery almost always end up squandering both the effort and goodwill of the local people. It is not clear whether Yan knows about Liao's failed handkerchief efforts in Daping when she says this, but Liao does not interrupt her. She continues, "Outsiders cannot easily replace their livelihood. We can only help them to their feet by helping them do better what they have been accustomed to doing for generations."

By the end of that day, Liao leaves the school triumphant, for she has made the principal agree to start offering special training classes for a few of Liao's own volunteers each year set to start in two years.

Dreams and Visions

The entire length of the road from the foothills to Daping Village is rutted, and its hairpin curves are edged by steep precipices. This is the road Liao helped build, and the first road that led to Daping village. The house of the Global Village (GV) volunteers stands with the rest of the more or less identical houses that form small clusters linked to the road at irregular intervals.

In my early morning run through the village to the mountaintop, the smoke of burning wood fills my lungs. For some reason, the energy-saving stoves designed to be part of the houses never became viable, not even in the Global Village's main house headquarters. However, I notice that the new village shrine is freshly painted, its incense smoke unspooling lazily from the altar.

The shrine is important because Liao's Lehe ritual calls for revering ancestors and tradition. The sense of peace and order is occasionally highlighted by the barking of the dogs that are friendly enough not to bite, as they often do in most Chinese villages.

At the end of the village road, I find a museum of sorts. The couple of traditional houses that had survived the earthquake stand limp, their walls sagging. A glassed poster-stand still holds what seems like the original posters containing the stories of these houses. But they are all smudged with mold. As it snakes up the hill, the road dwindles into a path of slippery mud, hindering my progress. I move slowly through the terraced *huanglian* fields scattered here and there. When I reach the top of the mountain, I find one lone man living in a twig shack that he built for himself after the earthquake killed his wife. When I told him that the GV volunteers were raising pigs in their Lehe project, the man responded, "Raising pigs? Won't work. The feed's too expensive."

Liao brushed aside what I relayed from the old man. The whole point of her new method, she says, is to reduce the grain in the feed by fermenting the added organic material. Her usually soft-spoken words could, on occasion, feel resolutely final or even biting. Such was the case, for example, when she commented on President Bush's call on people to shop more while standing on the site of 9/11 ruins. Such waste and consumerism are, to her, the extreme antithesis of nature preservation.

With only one full day in Daping, I continue to push open the doors that are never locked. And when the *huanglian* farmers start to trickle home at dusk and clean their muddy shoes on their covered porches, I find some who are eager to talk. From them, I discover another perspective on Liao's projects.

The tall sturdy man of one house, who preferred that his name not be used, is clear that he is not a fan of Liao's projects. He starts to tell me his opinions of Liao by saying her name in full, an act of disrespect according to Chinese custom. He says, "Liao Xiaoyi, we used to call her Liao Niang [mother], as we call other farm women, but now we call her Liao Laoniang [hag]. She has cheated us."

"What she does is no better than a scam." The man's grown son chimes in.

I learn that Liao's project has three times failed at raising organic pigs and chickens. A neighbor joins our conversation, explaining that GV had also leased prized fields to grow unfamiliar plants such as kiwi fruit. The result was no harvest and ruined land. Tourism has also been tried. Groups brought in by GV met all day in their guesthouse, but while the villagers did all the work in caring for and catering organic food for the tourists, they received a smaller share of the income than GV.

"She helped to build this road," the man says, pointing outside as he sits down with his family and some neighbors around a pan of charcoal fire, "but the money for our houses was from the Red Cross." He then adds: "To do more good? She isn't even here anymore."

I ask whether they are aware of what Liao's team has been doing lately. The neighbor says, "They can do all they want to kill time. We farmers need to eat. We can't run around doing their experiments. But the problem is that with their team here, nobody else can come in. We have no idea what she has arranged with the leaders. But there have been other interested parties who wanted to bring in business, and Liao Laoniang seems to be the only one sticking around. She's had four years and has nothing to show. Now she won't even give others a chance."

When I return to GV house, I discover that Liao was not pleased with my absence from her afternoon activities. She says she wanted me to follow her team's routine. But she nevertheless graciously invites me to join her for dinner at Secretary Yuan's home, where she proposes her future plans for Daping.

Later, while talking with some of the volunteers, I learn more about the operation. "Some of us are a bit afraid of her," whispers Shi Cong, a woman who will be my roommate for the night. To illustrate her point, she describes an incident when Liao had fired a volunteer on the spot for not being able to recite from memory the five aspects of Lehe Home. Another volunteer, Cao Yu, is more sympathetic with Liao and her hard work. "The farmers here have unrealistic hopes," he says. "Some people here expect us to hand-deliver to them a better life. Wuxi is better."

I discover that Wuxi County does indeed feel like a very different operation. Feiran, Liao's team leader in Wuxi, took me to Yangqiao from Wuxi, the county seat, a trip of about an hour's bus ride. From the bus station where we arrive, we walk through the villager-built Lehe Gate at the entrance of Yangqiao, the first site in Wuxi. Flat, wet rice paddies stretch down both sides of the narrow paved road where white ducks waddle and float, pecking at the stems left from the harvest.

The road splits in front of the village hall before continuing up to mostly new, two- and three-story homes on the hills. Music from a dance class on the second floor of the village hall drifts out from the open window, while children of elementary school age play in and around the building. Feiran tells me that the village hall functions as a free tutoring center on school days. Yangqiao and the other three sites each have two volunteers, usually a college graduate paired with a local volunteer with a secondary school education.

Taking a branch of the forked road, we come to a neighborhood where seven or eight elementary school boys and girls play in a yard while two grannies are chatting next to a nearby fruit and vegetable garden. They smile and nod as we approach. One chides the boy whose toy gun lands a shot on my leg as he squeals with delight at this accomplishment. This boy and his brother's parents are away working in the city, the granny explains to us apologetically. Thanks to the big sisters in the Lehe Home, she explains, the children at least have a good place to go from time to time.

Further down the road, the head of the village's Women's Association warmly greets us as we approach and invites us into her home. This auntie knows of all the work done by Lehe volunteers: home visits; weekly enrichment classes taught at the local elementary school; daily free tutoring; and weekend activities such as dancing, art, and reciting Confucius's classics. She says the most important thing is the fact that they are mainly there for the children, although they also help with adult education.

They taught villagers how to greet city visitors and shared parenting theories with young parents who are often around the volunteers' own ages. They also educate villagers on approaches to environmental protection. She gratefully adds that in a crunch they also help with the village's paperwork and bookkeeping.

Outside of auntie's house, the new homes stand as promises of a city job and give a glimpse of urban life. The walls ringing the fishponds display essentials of the Lehe philosophy in large characters: "To be grateful, to be content, to know pride and shame, and to know the etiquette; to respect heaven and earth, to respect you and me, to respect the myriad of beings that you see." As I walk down the narrow, newly paved road, I recall the young Daping mother who told me that high-rises and modern buildings are far more aesthetically pleasing to her. Daping must have looked similar before the earthquake destroyed the modern cement buildings, I thought. Yet the newly constructed traditional-style homes redefined its landscape, perhaps making it a unique new village in China's hinterland.

Most of these new homes in Yangqiao seemed empty, and Feiran comments ruefully that the village is losing its soul, just like many others across the country. Young people have left for the cities, she says, and they and their children will never know how to farm. Despite her feelings about the people losing their farming traditions, however, Feiran herself has abandoned the past. A 24-year-old college graduate, she has been in charge of Liao's Wuxi team for over a year. She says she admires Liao for her fight to strengthen the community in the village, and that she promised Liao to stay for at least three years. "But," she adds, "I'm not sure I can afford to sacrifice my youth to this." She hopes to go to graduate school and do some thinking and learning in search of her own path.

Feiran was about the same age as Liao's daughter Karla, who was, in many ways, already living the life Feiran aspires to. But Karla has a mother who gave her *Sophie's World* as her tenth birthday gift and single-handedly raised her to become another Sophie.[14] Karla was already a student of philosophy and mass media at an Ivy League university in the United States. In a long online debate on Eastern and Western philosophy that Liao forwarded to me, Karla pressed her mother to delineate the Eastern philosophy she believed had inspired her work. Frustrated by her mother's abstract language, Karla resolved to attribute her mother's action to her personality rather than to the philosophy: "The difference between us, is one between a saint and a non-saint. . . . The reason you haven't jumped out of the window is because you are a compassionate person. That's why you're not giving up."

Where the Future Meets the Past

To Liao, Karla was the brightest spot in her life. During Karla's first year at college in 2011, Liao said she watched the Chinese New Year program on TV by herself for the first time. When she heard the children of migrant workers sing, "If you love me, then accompany me, kiss me, praise me, and hug me," her heart ached for the absence of her daughter, but she was also saddened by the theme of urbanization where workers and farmers rejoice about moving into urban apartments which Liao calls "cement boxes."

That night, Liao wrote the following in her diary: "Regardless how many reasons they have to leave children behind, I still cannot understand those migrant parents who do not return to see their children even in Spring Festival time. Because separating from my child is one single torture I, the so-called strong woman, cannot endure." Memories filled her pages that night as she wrote. Liao recalled how the debilitating separation from the not-yet-four-year-old Karla had almost thwarted her resolve to finish her first film in China, and how Karla had been with her at every step of the way as she blazed a trail of environmental protection in her country. As she searched around the world for an answer—from the muddy tent in Daping to the celebratory striking of Yangqiao's ceremonial bell—her daughter had been there, with a sweet kiss every morning. "For 22 years, I have not tagged along with her, but she has tagged along with me!" Liao wrote in her diary.

Liao's rural campaign aims to restore the sense of home and improve the quality of life of both urbanites and villagers. In her vision, she sees organic food and a clean environment for city-dwellers and for villagers, reunited families, and a return to traditional communities. She hopes modern transportation and communication systems will connect these two worlds.

In Daping, Liao has tried to bring back migrant workers from the cities with the promise of more profitable organic farming. In Wuxi, her team's main endeavor now is to care for migrant workers' children and to keep the community together. The results of such endeavors are uneven to date. Having done some of the proposed clean-up work around Yangqiao village, Lehe's other, more ambitious projects have either been forgotten or lumped in with other unfulfilled government promises that so often fade over time. Some of the limited success cases include the few mu (1.65 acres) of experimental organic fields still in operation. The rice that they have yielded, Feiran said with relief, has just been sold.

Liao is clearly aware of the situation, but she explains that the results of these social experiments will take time to manifest themselves in a manner analogous to the work of Chinese medicine. For each failing effort, she can point out specific reasons, and furthermore, is able to offer many stories showing the promise of her approach. For example, her Lehe team helped raise the Beimengou villagers' awareness of the harmful impacts of a particular pile of accumulated garbage, and advised them to take ownership of the problem. These villagers then worked with leaders and banded together to remove the pile. As it turned out, the entire project of garbage removal was not only completely resolved after years of environmental pollution, but also saved government funds that would have gone to removal costs.

Many such successes and moving stories told by Liao have found their way into "The Way of Sustained Stability and Long-Term Security," a 2012 report written by Dr. Pan Wei and Dr. Shang Ying and published by the Center for Chinese and Global Affairs at Peking University.[15] Yet the report makes no distinction between discontinued practices and those that are ongoing. The truth is that many of the Lehe practices have been hard to quantify. Even Feiran says she is not certain, for example, whether her team's work would slow down or increase the outflow of migrant workers. She points out that if parents feel their children are being cared for, they might be more inclined to stay away from home. Only time and numbers will provide more definitive answers.

Indeed, difficulties remain for Liao's social experiment in the countryside as she moves forward. Getting enough funding is a perpetual challenge for any NGO, even with a prominent leader such as Liao. It is also difficult for her to recruit and keep volunteers in an increasingly commercialized society. There are also potential political problems. Because her Lehe work is so dependent on local government, every change in local leadership threatens the stability of the partnership, and thus the effectiveness of her work. And finally, but not the least of her struggles going forward, she faces challenges from some of the very local people she is trying to help. Despite these difficulties and uncertainties in her Lehe projects, Liao's earlier environmental protection work along the lines of traditional Western practice has become the foundation of many current practices in Beijing, cementing her status as a celebrated environmentalist.

In 2000, the Sophie Foundation chose Liao as their third (and first Chinese) recipient of its prestigious Sophie Award for this reason: "Twenty-two percent of the world's population lives in China. As China prospers through an unprecedented rate of economic growth, the choices China makes when it comes to environmental resource management will define the future for all humankind."[16] At the Banksia Award ceremony the following year, Liao proudly said, "Rachel Carson also got the same award. It means I did something right and that anyone can make a difference."[17] In 2008, the Global Citizen's Award was presented to her by Bill Clinton for "commitment to leadership in civic society."[18] In 2009, Liao was named a Hero of the Environment by *Time* magazine.[19]

"What is the price of growth?" Liao asked at her featured speech at the World Affairs Counsel in Oregon in 2007. "We may have less poverty, but we have more pollution. Today China suffers, tomorrow the world. We need to work together."[20]

At a time when much between China and the West is in

dispute and, in particular, when there is disagreement about how to tackle the environmental degradation resulting from China's rapid industrialization, Liao and her work remain one of the few things the two parties can agree to promote.

Asked if the Kyoto Protocol needed revision, Liao said, "Yes, but according to different environmental responsibilities. We do what we can, not just what we want to do."

On her relationship with the government, she says, "An NGO is small, government is powerful. Partnership is necessary to make an impact."

Although she detests being "wasteful," a term she associates with consumerism, she is nevertheless philosophical about it: "If the whole world lived a wasteful lifestyle, how can you just ask Chinese to conserve? We're living in the same world. Everyone needs to change. That's the only way we can save the planet and ourselves."

Many of Liao's goals are in line with the Chinese government's. Harmony is the most prominent of all. To bring stability and harmony to society, to return city-dwelling migrant workers to their rural homes, to care for these workers' children, to improve the financial condition of the rural poor . . . all of these ideals are high on the government's agenda. Looking ahead, Liao is confident. Her latest victory is a commitment from the United Front Work Department of the Central Government to promulgate her program of caring for children of migrant workers left behind in villages.[21] "Go spread the word to the English-speaking world," Liao tells me as we part in the early morning darkness in Daping village, "that another lifestyle is possible."

As a disciple of Rachel Carson, Liao has come full circle from being an admirer of a market economy to turning her back on Western industrialization and consumer culture. Now she wants to export her version of Chinese-style environmental protection back to America and the rest of the industrial world. In the face of China's rapid industrialization, urbanization, and commercialization, Liao is now fighting an uphill battle. That her results have been mixed underscores the complex interaction in contemporary China between state and citizen, educated elites and ordinary people, and conflicting aspirations for economic growth and the need for clean water, clean air, and a safe food supply. But ultimately, Liao says, people can, and should, live in harmony with nature in order to save their common home, the Earth.

Chapter 1 Notes

1. Wenchuan Earthquake, according to the Chinese government data, measured at 8.0 Ms and 8.3 Mw (the US measurement being 7.9 Mw), and occurred at 2.28 p.m. Beijing time on May 12, 2008, in Sichuan province. Chinese Wikipedia puts the numbers, by noon of September 8, 2008, at 69,227 people dead, 374,643 wounded, and 17,923 missing. Fifteen million people lived in the affected area. The Western media estimated that the earthquake left about 4.8 million people homeless. On the 2009 anniversary, CCTV announced the official numbers as: 68,712 dead, 17,912 missing (http://news.cctv.com/china/20090507/108604.shtml). The National Earthquake Administration posted a different set of numbers at noon on June 22, 2008: 69,181 dead, 374,171 wounded, and 18,522 missing (http://www.cea.gov.cn/manage/html/8a8587881632fa5c0116674a018300cf/_content/08_06/22/1214124901775.html). There has not been a definitive set of numbers to date.

2. This and other local numbers in the following were provided by Liao Xiaoyi and corroborated by the local production team leader, Mr. Yuan.

3. I first met Liao Xiaoyi in September 2012, and my research proposal to study her work was warmly received. From the fall of 2012 to the spring of 2013, I had two in-depth interviews with her, took a week-long trip with her to Sichuan to visit her Lehe villages in November 2012, and attended many of her activities, mostly lectures and discussions in her headquarters, as well as some social occasions. Among them was a series of talks and discussions on the relevance of Liang Shuming's work in the new era. One such talk featured the author of *The Last Confucian*, Professor Guy Alitto of the University of Chicago, who was in Beijing for a conference. In our frequent email and phone exchanges during this long period, Liao also shared with me transcripts of some conversations to help me understand her work.

The volunteers in Liao's projects are a notable group. The team members rotated every few weeks from site to site, allowing them to get to know each other at these talks and celebrations where I was often invited to participate. Volunteering at NGOs at the time was a rather new concept for young Chinese college graduates and, in many ways, an altruistic pursuit compared to the more trendy drive for academic degrees and high income. These young volunteers I have come to know on many very different occasions are often idealistic. Despite their often humble backgrounds, they have decided to go against their parents' advice to get better-paying jobs. Several times I heard them say that they wanted to see if there were things beyond money, and they wonder what those things might be.

4.   Rachel Carson is the author of *Silent Spring* (1962), a book that exposed the harm done to the environment by chemicals, and spurred a reversal in national pesticide policy. *The New York Times Book Review* of that year published an article titled: "There's Poison All Around Us Now." The book inspired an environmental movement that created the Environmental Protection Agency.

Liang Shuming (1893-1988) is a philosopher and the author of many books. In his best known book, *Eastern and Western Cultures and Their Philosophies* 东西文化及其哲学 (first edition published in 1922; reprinted in 2006 by Shanghai People's Press 上海人民出版社), he compared Eastern and Western philosophies, and came to the conclusion that Western methods would never suit China. Believing that the Chinese countryside was the repository of traditional Chinese values, he helped found the Shandong Rural Reconstruction Research Institute. The Japanese invasion of 1937 effectively ended this endeavor. For a comprehensive study of Liang, see Guy Alitto, *The Last Confucian: Liang Shu-Ming and the Chinese Dilemma of Modernity* (Center for Chinese Studies, UC Berkeley, Book 20; 2nd edition, University of California Press, 1986).

Suolandajie (1954-1994) was a Tibetan and an "Environment Defender" (a title awarded by the Chinese Ministry of the Environment). He died fighting a group of poachers who had killed a great number of endangered Tibetan antelopes in Kekexili, a vast protected region in Qinghai province.

5.   These are two prestigious international awards for environmental work. Liao calls the Sophie Award "The Nobel for Environmental Work."

6.   Published by *SunChime*, 2010.

7.   Jieqi 气 refers to the traditional 24 divisions of the year. Each period (*jieqi*) is characterized by different features of the climate that would guide agricultural activities in different seasons. *Yangshengcao* 生操 literally means "nurturing life exercise."

8.   See Guy Alitto, *The Last Confucian: Liang Shu-Ming and the Chinese Dilemma of Modernity* (Center for Chinese Studies, UC Berkeley, Book 20; 2nd edition, University of California Press, 1986).

9.   Yingchun Wang, "Lehe Home: An Idealist's Village Building Utopia," *Journal of Chinese Economy* (October 27, 2010).

10. This information was collected from talks with both volunteers and villagers, as is all other information related to her failed attempts in Daping.

11. The First Social Service Microblogging Competition 首届公益微博大 was held in a square in the city of Chengdu on November 20, 2012. It gathered hundreds of people who worked in mostly small NGOs performing various public services. I have met and talked to volunteers who drove mobile libraries to remote village schools, caretakers for the aging, and many others. They all competed in blogging to expand their influence. Posters lined the

conference site, describing the nature and work of each group. This conference was held to choose the most effective of these groups in their blogging efforts.

12. Part of the lyric of this song, "Firefly," is translated by the author as follows:

萤火虫，萤火虫，慢慢飞	Firefly, firefly, fly slow
夏夜里，夏夜里，风吹	Summer night, summer night, the breeze gently blows
怕黑的孩子安心睡吧	May the dark-fearing child sleep tight
萤火虫你一点光	As the firefly will give you some light
燃小小的身影在夜晚，	The small body burns in the night
夜路的旅人照亮方向	Shining the way for the travelers in the dark
短的生命，努力的发光	In this brief life, you burn bright
黑暗的世界充希望	Filling the dark world with hope and light.

13 . The full name of this college is Polus International College 四川国际榜业学院. This is a vocational school established in 1993. It quickly gained national fame after having been chosen to assist with the production of some TV series, as well as the make-up work for the 2008 Olympics performers.

14. *Sophie's World* is a 1991 novel by Norwegian writer Jostein Gaarder. The protagonist Sophie Amundsen is a 14-year-old girl living in Norway. The story is about how Sophie was introduced to philosophical thinking and the history of philosophy by a 50-year-old philosopher named Alberto Knox.

15. Dr. Pan Wei 潘 received a PhD in Political Science from UC Berkeley in 1996. He is a professor at the School of International Studies. Dr. Shang Ying 尚英is a colleague of his with a PhD in Political Science from Harvard. Dr. Pan is one of the scholars who are often referred to as the "new left." He is known as a fierce proponent of one-party rule. For related discussion, see Suisheng Zhao, ed., *Debating Political Reform in China: Rule of Law vs. Democratization* (M. E. Sharpe, Inc., 2006).

16. See Jury's Statement for the prize at http://www.sofieprisen.no/Prize_Winners/2000/index.html.

17. The speech was shown to me in Liao's office, and I do not have a copy. An online search in October 2015 on the Banksia Environmental Foundation did not turn up a video record.

18. See the link at http://www.clintonglobalinitiative.org ourmeetings/2008/meeting_annual_GCAwards.asp?Section=OurMeetings&PageTitle=Global%20Citizen%20Awards%202008 "Annual Meeting 2008: Global Citizen Awards." *Clinton Global Initiative* (accessed March 29, 2016).

19. This can be found at http://content.time.com/time/specials/

packages/0,28757,1924149,00.html "Heroes of the Environment 2009," *Time* (March 29, 2016).

20. The video can be found at this site: http://library.fora.tv/2007/05/29/ This_Endangered_Planet_A_Chinese_View. World Affairs Council: Oregon. "This Endangered Planet: A Chinese View." *FOR A.tv.* (accessed March 29, 2016).

21. Liao informed me of this verbally.

CHAPTER 2

A Sexual Revolution with Chinese Characteristics: Li Yinhe's Campaign for a More Open Society

At the spring 2013 "Women's Media Awards" ceremony sponsored by UN Women and 163.com, Li Yinhe walked on stage as an award-presenter. Cheers and some floor stomping swept through the hall in recognition of China's one and only female sexologist, an American-trained sociologist recognized by *Asiaweek* in 1999 as one of the "50 Most Influential People" in Asia.[1] Li's round face framed in short hair was projected onto the large screen behind her. Her smile at the welcoming audience had a touch of bemusement, perhaps because of what might have seemed to her a misplaced enthusiasm directed towards the presenter rather than the presentee not yet on stage. She nodded in acknowledgement as she approached the podium in unhurried steps, giving time for the cheers to subside. When the ceremony was over, friends and admirers alike came up to greet her. "Teacher Li" they called her, and she alternated between warm greetings and assertive statements when asked to comment on a topic. At one point, she said, "Among the nine members of the standing committee of the Politburo, none is female. This shows great gender inequality in our country."

From the fall of 2012 to the spring of 2013, I had the opportunity to interview this fascinating and controversial woman a few times. During these interviews, we talked about the topics that were uppermost in her mind, including the study of sex and gender, the feminist movement, Chinese family structures, philosophers, and politics. But time and time again, we circled back to some key issues high on her agenda: same-sex marriage and prostitution. These are topics of public contention in China and the United States. But to Li Yinhe, the debate in China is primarily about the individual's place in an increasingly open society.

Li Yinhe is a complex figure. Although highly regarded by many, she is also seen by some as an apologist for an immoral lifestyle heavily influenced by Western values. Some in this latter group even use her name to make their point. Literally translated, "Yinhe" means "Silver River," the Chinese name for the Milky Way. Yet some of her online detractors switch the character "silver" to one that means "licentiousness," which is pronounced the same way, thus suggesting their vision of her as an advocate of indecency. Yet this polarizing figure is actually soft-spoken and smiles easily. Her warm and gentle personality often contrasts with her fearless battling words in the media.

In the time leading up to the highly anticipated 2013 annual conference of the National People's Congress (NPC) and the Chinese People's Political Consultative Conference (CPPCC), the topic of discussion was more focused on politics than sex for this sexologist. Li Yinhe has a special connection to the annual conference of the NPC, China's national legislative body. Almost every other year over the past ten years, she had submitted her proposal to legalize same-sex marriage. Due to a lack of support among the delegates, however, the proposal has never become a bill. But in 2012, Li scored a victory by gaining the support of the parents of gays and lesbians who submitted a letter of support along with her proposal.[2] "I will continue to submit this until it becomes law one day," Li said to me. "Legislation always lags behind the times a bit, but it will eventually catch up." For her, the legalization of same-sex marriage is a matter of fairness, equality, and justice: to extend equal rights to those with different sexual orientations is the right thing to do. As she said, "If we see something that's not right, we should correct it."

Thirty years ago, at the inception of Reform and Opening,

submission of a formal petition to legalize same-sex marriage and getting parental support for the effort was unimaginable. Prior to this time, people received capital punishment for extramarital sexual liaisons that were merely alleged.[3] Homosexuality was not even a topic for discussion. For reasons such as these, China has not been known as a respecter of individual dignity, and its authoritarian government has often been cited in the West for cultural and political repression. If China is to present itself as a modern and civilized society, this will have to change.

In the course of the past thirty years of Reform and Opening, the government's reaction to extramarital relationships has changed from the most severe punishment to decriminalization.[4] And Li Yinhe has been the most outspoken public intellectual in the campaign for such change, often by staging a one-woman show that keeps discussions of sex and sexuality in the public debate. With over 40 books (just to date) online[5], blogs, and speeches in public forums, she speaks on behalf of people's most private, most unspeakable, and thus most vulnerable issues. She calls for government and society to accord equal dignity to those with different lifestyles or sexual orientations, and she also calls for individuals to speak out and fight for their own personal rights and freedom. Li makes sure that respect for the individual's voice and self-expression are part of the reform in building a modern civil society. The campaign of China's only female sexologist offers a glimpse into the complexities of the country's struggle to balance tradition and individual rights, freedom, and dignity.

In many ways, Li is an unlikely fighter for the rights of the lesbian, gay, bisexual and transgender community LGBT, for she has always been an exemplary figure in society. As a student, she sailed through school with ease and distinction; at age 16, at the height of the Cultural Revolution in the late '60s, she volunteered to answer Chairman Mao's calls to go to the countryside to be re-educated by the peasants. While in the countryside, she decided again to go to the most different and remote place she knew: Inner Mongolia. So she wrote her petition letter in blood, and got her wish. Today, like many in her generation, she credits that period of hardship for toughening her after an earlier life as a city girl. By the time the college entrance examination was reinstated in 1977, she had already finished college. That year, she got a job as a government newspaper editor, the same profession as her parents, and she met her now late husband, Wang Xiaobo, a writer still idolized by many in China.

In fact, despite her own achievements and fame, she is still sometimes referred to in an occasional footnote as "Wang Xiaobo's wife." This is perhaps more due to the fairy tale love between the two rather than the need to offer her more identifying features. Li Yinhe published many of their love letters in a book called *Loving You is Like Loving Life* after Wang's premature death in 1997 at the age of 45.[6] Recently retired at age 61, Li often reminisces about their 20 years together, invariably starting by telling of how Wang helped start her study of homosexuality.

She came to the topic by chance. In 1989, freshly back to China with a sociology PhD from the University of Pittsburgh, she accepted a post-doctoral position at Beijing University with China's most famous sociologist, Fei Xiaotong. Quoting Fei's mantra, she said, "Sociology should produce stories because society is a stage where human stories unfold." One of Li's first social projects was to interview people in Beijing who had chosen to remain single. "You'd be surprised how difficult it was to find such people in China," she said. "Among the over 40 people who answered my advertisement on the Beijing Evening News, there was one man who, when asked why he remained single, said to me, 'You seem like a nice person, so let me tell you the truth: it's because I'm gay.' From this it snowballed to over 120 people." Because some gay men were uncomfortable talking to Li, she and her husband coauthored their first and only book together, *Their World: Looking Into the World of the Chinese Gay Community.* It was published in Hong Kong in 1992.[7]

Sex, the Individual, and Social Change

On a drizzly spring day in Beijing, we meet in her suburban Beijing villa. Wearing a casual grey warm-up suit, she waits for me at the gate after guiding my taxi driver to her place by phone. A smile of relief spreads over her pale face as I arrive. She is without her customary eyeglasses, but the lower rims of the glasses seemed to have etched a pair of permanent curves below her eyes. Her voice is gentle and relaxed, as if she is greeting an old friend. We walk toward the house through her neatly arrayed garden. Her head is held high with an air of defiance against the pull of gravity. Her steps are soft but sure.

Once we are in her personal study adjacent to her bedroom on the second floor, she settles into a midnight blue velvet chaise lounge. We talk about all the topics she is passionate about, and there are many: homosexuality; the upcoming LGBT conference in Salzburg that she will attend the following week; pornography; polygamous relationships; premarital sex; the feminist movement; law; traditional Chinese family structures; traditional Western studies of sex; and her favorite philosopher, Michel Foucault. Age and poor health have softened and slowed her somewhat, yet her well-trained mind leaps and bounces through times and cultures with agility and passion.

Li then turned to two topics high on her list at the moment: same-sex marriage and prostitution. For her, both issues are about human dignity and the individual's right to their own chosen lifestyle. This is a battle on two fronts: Chinese tradition and the legal community. And the progress of one hinges on that of the other. To legalize same-sex marriage, for example, will firmly put individual rights over fundamental social obligations mandated by tradition. That is, to propagate the family is an individual's primary duty.[8] To accomplish this, she has decided it is necessary to work with the government and its legal system in order to bend laws away from the established social order and towards the individual.[9]

In her polls and studies, Li sees great potential for social progress. To begin with, her 2008 poll shows that China already has a surprisingly high tolerance for homosexuality. Of the 400 random respondents in Beijing, 91% believe gays and lesbians have equal rights to employment. This is much higher than the 65% shown in the United States in a 1983 poll, 86% in 1996, and even 89% in 2008, the same year Li's data was collected.

But there is work on other fronts. For example, only 20% of her urban respondents regard homosexual relationships as completely acceptable, 30% think that they are somewhat acceptable, close to 40% regard them as completely unacceptable, and close to 10% have no opinion. This reflects a more ambivalent view compared to the American numbers that Li quoted: 43% complete acceptance and 47% complete rejection of homosexual relationships.[10]

When it comes to the issue of the legalization of same-sex marriage, Li quotes recent polls conducted by phone which reveal that 27% of Chinese citizens favor legalization while 70% are against it. But Li points out that online polls in the same period show that twice as many—60% of Chinese netizens—are for legalization. The fact that the younger and better-educated netizens are more sympathetic to and supportive of the LGBT community's equal rights gives Li confidence in continued social progress. Still, she concedes that Chinese gay-marriage opponents at 70% are significantly more than the 58% in America, even though this is better than 83% in Hong Kong; and likewise, while the 27% who support this in China are also significantly lower than 40% in America and higher than the 17% in Hong Kong.[11]

"If society changes, its customs should follow." Li pauses, as if reminding herself: "Customs have strong inertia. But even they can be changed. Otherwise, Chinese women would still have bound feet, because this was the custom. Just because traditionalists haven't felt the pain the custom has inflicted on women doesn't mean that women themselves can't do something about it." In her work, Li affirms the power of humans to change, steadfastly stands with individuals in society, and urges that rules and regulations, as well as age-old traditions, ought to change in sync with the times instead of confining and restricting human progress. She chides and criticizes the traditionalists' obstructions, and explains conflicts and difficulties through her own studies. In her first book, *A Brief History of the May Fourth Movement*, Li examines the landmark event of 1919. "Love is new to China," she says, referring to sexual love and its dawning as a topic of open discourse in society. "What we had in our tradition was arranged marriage for the sake of continuing the family line. In the 1919 May Fourth Movement, young people found love, and they eloped."

In a string of subsequent social changes, this rebellious spirit and many of these youths were part of the driving force that propelled the Communist Party to power in 1949. Li's own parents were among those who joined the party and rose to leadership positions. And it was also this rebellious spirit that inspired the CCP to draft a new marriage law as the first law of the land after the constitution was written. This law enshrines marriage for love, as well as the equality of women. But over the next 30 years, relentless revolutions swept through the country. Although women were urged to take "half of the sky" as equals of men, individuality was vehemently opposed.[12] She laments that generations had come of age and were unable to identify their own personal desires and preferences—her own mother included—because their personal compasses had been set to uphold the majority interests.

As a complete break from that revolutionary past, the paramount leader Deng Xiaoping initiated Reform and Opening in 1978. Accompanying the explosive wave of economic and social change that Deng unleashed was a tortuous path of legal reform in an attempt to regulate individuals' social behavior. The first set of criminal laws drafted in 1979 was undertaken after the Western model. But the earliest version of the criminal law included an ill-defined crime called "hooliganism," as Li points out. It basically punished all extramarital sexual conduct, including homosexual relationships and implied or suspected sexual liaisons. This legal item was generally viewed as largely responsible for what is now known as the "Strike Hard" campaign of 1982–83. In that campaign, budding heterosexual behavior—completely innocuous in today's China—was punished with a severity that is almost inconceivable today. Li Yinhe personally knew of people who received capital punishment for merely being accused of having extramarital sexual relationships.

Some of the most publicized cases at the height of this

campaign testify to the power of tradition when backed by new-found legal expression. Li recalled the 1984 case of Chi Zhiqiang, one of the most popular movie stars at the time. While he was detained for allegedly improper relationships with women, two journalists coerced him into revealing details of his story in the detention center with a false promise of lightening his sentence. The account published in newspapers caused such public outrage that despite the intervention of the premier's wife at the time, Deng Yingchao, Chi was given four years in prison, which was eventually reduced by half for his exemplary behavior while incarcerated.[13] Two other men who happened to have been sentenced with him on the same day were not so fortunate: one allegedly peeked into women's public bathrooms multiple times, for which he received a death sentence deferred by two years; the other was accused of forcing an embrace on a young woman, and he was sentenced to four years in prison.

Bending the Law Toward the Individual

At the time of Chi's conviction, Li Yinhe was already in Pittsburgh. Even from afar, Li could see the power of law in an authoritarian society. She compared it to the official edicts of the past that the government used to announce and then implement policies. The criminal law in its infancy in the early 1980s, in Li's view, served the traditionalists like an executioner's knife to defend traditional moral standards. Chi is now in his fifties and, according to occasional evidence in the media, he still lives with the hooligan stigma. "If I had been born 20 years later, I would not have been sent to jail!" he is quoted as saying.[14] Li comments to me that the austere environment of the early 1980s when basic sexual expression met with such severe punishment meant that issues surrounding homosexuality had no chance of coming to light.

Li points out that in Chi Zhiqiang's case the newly drafted criminal law in its earliest stage put two opposing forces in sharp focus: traditional values and an increasingly liberal society. The rule of law being still new and uncertain, traditional values held sway. Looking down the road, Li feels that this conflict between modernization and tradition will continue to play out in China's subsequent modernization process. Because the laws are there to stay, the legal profession and law enforcement will learn on the job that people's needs for sexual expression will not only remain but will become more clearly manifested as they are fueled by increasing awareness of how Western cultures focus on individuality and individual expression.

Much progress has been made since those initial years of the criminal law. With hindsight, Li is willing to give the government credit for its attempt to adapt. She easily lists the milestones of the legal relaxation, which to her are a measure of the gradual liberalization of Chinese society: In the first major revision in 1997, the crime of "hooliganism" was removed from the criminal law.[15] This effectively decriminalized all sexual activities between two consenting adults in private, including extramarital and homosexual sex. "Hooliganism" was replaced with the more narrowly defined crime of "assembling a crowd to engage in promiscuous activities." This new item basically punished group sex. As the legal code emphasized, the victim of such behavior is the established social order.[16]

The crime of "assembling a crowd to engage in promiscuous activities" was later moved from criminal law to the newly revised 2006 "Public Security Administration Punishment Regulations of the People's Republic of China," which carries a lesser punishment. This law came into effect in 1987 and was replaced in 2006 by a new set of regulations under a very similar name.[17] The punishment in the new regulation carries a sentence of 10–15 days of detention or a fine of 500–1,000 yuan (about $90–170 USD).

Homosexuality is a different issue. When the Chinese Psychiatric Association removed homosexuality from the official list psychiatric illnesses in 2001—almost 30 years later than the U.S. (1973) and eight years later than the World Health Organization (1993)—homosexuality was finally free from officially sanctioned discrimination.[18]

In Li's view, traditional morality is one constant obstacle to the law's keeping pace with people's increasingly open sexual practices. In media interviews and her own articles and blogs, Li constantly engages conservatives in debate. One of her better-known critics, for example, is a law professor named Lu Ying. When Lu echoed the common conservative refrain that "Law maintains the baseline of morality," Li retorted that morality was exactly what she wants to separate from law. She argues that law should not be used to judge or punish moral issues because the moral baseline could be fuzzy and its perimeters are often made arbitrary by tradition. One case that Li likes to quote is the 1984 landmark American case that dealt with the Dworkin-MacKinnon Anti-Pornography Civil Rights Ordinance. In that case, city officials in Minneapolis cited the freedom of speech guaranteed by the First Amendment in order to strike down the argument that pornography violated women's civil rights.[19] By using this case, Li tries to illustrate that law should safeguard individuals' rights and not enforce a moral standard. In a country such as China where law is traditionally used by the ruling party to maintain social order, Li wants to bend it toward safeguarding individual rights, as is practiced in the States.

Law can be harsh, Li says, but it also has the power to change life for the better. Li sees this power in her friend Cui Zi'en's ordeal. When Cui openly announced he is gay to his class at the Beijing Film Institute in 1991, he was immediately dismissed from his position, banned from teaching for ten years, removed from his dorm, and ordered to enter a hospital and fix his sexual orientation. While living as an outcast in those ten years following his announcement, Cui worked furiously to write novels and film scripts that speak out for the gay community. Only when homosexuality was officially eliminated as a crime (1997) and as an illness (2001) did Cui finally regain his previous teaching post.

"While Westerners tend to explore what can be done outside of the law, Chinese like to stick to what is prescribed by the law," Li says. This awareness of the Chinese attitude towards law leads Li to keep pushing for legal reforms. She needs law to be a tool and the government a partner, however reluctant she may be in taking this approach.

When Li participated as an expert advisor in the government's marriage law revision of 2000, she proposed that the committee legalize same-sex marriage and prostitution. But the committee told her that "our society is not yet ready for that." Over the years, Li has indeed come to face serious pushback from the legal community, social conservatives, and makers and enforcers of government regulations.

Decouple Chastity and Virtue

Li understands that hers is a society that has long repressed sexual expression. In a tradition that says "for women, having no talent is a virtue," she sees how keeping women ignorant leads to a blind faith that is extoled as virtue in a male-dominated society. Chastity is one example of such a practice. She points out that no other government in history has, like the Chinese government, erected monuments to women who chose chastity over their own lives, either choosing to die with their husbands or to remain chaste at any cost after having been widowed, even at a very young age. "Here's today's society," Li says, reciting from memory: "In 1989, those who reported having pre-marital sex were 15% of the population, and in 2011, the percentage shot up to 71%. Society has changed. You have postponed marrying age from puberty in the old days to now 20 for women and 22 for men respectively, you have separated sex from making babies for the family by limiting the couple to one child, and when the sterilization rate is as high as 90% in some areas, premarital and extramarital sex are sure to rise. Because sex is fundamentally an activity of pleasure."[20]

To break women out of the cycle of aspiring to such "virtue," Li writes relentlessly about the bondage of women in traditional society. From 1991 to 2000, she published 17 books on women and sexuality. But in her criticism, she takes a page from Lu Xun, China's most celebrated writer of the 1930s. His short stories mercilessly expose the ignorance and foolishness of the Chinese, sometimes with biting satire. One story Li quotes is "Medicine," in which Lu Xun describes a father who, eager to save the life of his only child from an illness in order to continue the family name, follows a local doctor's advice to procure, as medicine, a steamed bun dipped in human blood. When a condemned revolutionary is paraded down his street one day, he follows the convict to the execution ground for the bun soaked with the blood of the executed revolutionary martyr. Naturally, the medicine fails, making Lu Xun's satiric point that tradition does not have all the answers, and that blindly following it is foolish.

Li says she sees such "medicine" in many forms everywhere. For example, a Chinese women's association rewarded exemplary women with the free installation of a new hymen, and a 38-year-old Wuhan woman posted her proof of virginity online as a call for defending chastity.[21] "A 38-year-old virgin is a joke." Li chuckled dismissively in her mild manner after she told me the story. Commenting on the poll she conducted in the 1980s, she explains: "In China, 28% of women never knew, as the poll showed, what an orgasm is, compared to 10% in the rest of the world; and 80% of women did not know that the clitoris is the most important part of the body to achieve orgasm. This is astonishingly ignorant."[22]

She encountered such ignorance early on. In her high school, for instance, she told her puzzled classmates that eunuchs were "castrated men" and was ostracized for knowing too much about both sex and men. The new China may have liberated women in Li's view, but the relentless suppression of the individual in the PRC's first 30 years of successive revolutions made sex and sexual expression a matter to avoid. Consequently, generations have grown up ignorant of the most basic knowledge of sex.

"But a more poignant point is another poll asking women which is more important, life or chastity," she says. "The result shows that 70% regarded chastity as more important. Now, you have equal numbers of people in both the chastity camp and those who practice premarital sex. What does that say to you?"[23] To answer her own question, she explains that people's deep-rooted traditional values are answering the chastity question while their real life needs dictate their actions. Therefore, a law that decriminalizes extramarital sex would free people from the guilt imposed by the tradition as they make their life choices. "That said, I'm not advocating extramarital sex, as many have accused me of." Li is very careful to make this distinction: "Advocacy for the basic freedom to make personal choices and promoting the practices are different things. All I'm saying is to give people room to make their own choices. And if these choices do not harm others— no matter how distasteful they may be to some people—people should be entitled to do what they want in private."

Li stumbled into the vortex of a confluence of tradition and changing social mores in 2006. This very first personal encounter left her with an acute sensitivity to the distinction between advocacy for rights and the promotion of specific acts, as well as an appreciation for the formidable power of tradition. It was at the invitation of the Jiangsu TV Station that Li gave a public speech on love in Nanjing on July 21, the seventh day in the seventh month in the lunar calendar that year, a traditional holiday for lovers. In her talk, she called for the decriminalization of all sexual activities as long as they met three basic conditions: being private, having no victim, and involving consenting adults. All she did was point out a new trend in society and the need to understand and tolerate personal sexual desires and expression. Yet, she recalls, "news started to circulate that I was promoting the things I thought should be decriminalized: same-sex marriage, prostitution, spouse-swapping, polyamorous relationships, sex orgies, and so on. I mentioned that, to many, these activities may be in poor taste, but they are within people's rights as long as they don't bother or harm anybody. So they should not be punished by law." The local newspaper, *Jinling Evening News,* printed a provocative headline for its report of this event: "Li Yinhe Promotes Polyamorous Relationships: Avant-Garde View Causes Mass Anger."[24]

"All the 'mass anger' they were talking about was just one older couple among many questioners in the Q and A section, cautioning me to be a bit more prudent when talking about sexual relationships," Li calmly explains. But judging by her blog entries, she was much more emotional in the wake of the storm. She wrote: "Oh, Chinese, I'm speaking out for your most vulnerable private thoughts and desires! And the response I get has been so hysterical! ... As citizens, you have the right to do what you want with your own body, this is our most fundamental right!"[25] But, she says with a sigh, "The more I explained, the worse it got." She smiles wanly as she looks back. The accusations and exchanges on her blog became so personal and toxic that she shut down her site's "comments" window three separate times. Eventually, however, she decided to leave the window open as a forum for people to work out their differences. Letters flew in also, some cheering for her, but many threatening her, blaming her, and trying to persuade her to move away from her views. "The attackers wanted to burn me to death in the square," she says, invoking Giordano Bruno's burning at the stake for what the Venetian Inquisition regarded as the heresy of calling the sun a star. "People tend to confuse moral standards with legal standards," she explains.

But her supporters did much to offset these attacks. She adds, "This is really a great honor for me. The admirers told me that I was their lighthouse. And a prominent Australian judge told me at a conference that I was his hero." The lighthouse metaphor came from a young woman in a lesbian bar after one of her speeches. Li was later received a model of a lighthouse that is now enshrined in her home.

The lighthouse is her bond with the LGBT community, a community she views as one that refuses to compromise with society's traditional forces. She says she was moved by the sentiments of one user of *Danlan*, or *Pale Blue*, the popular gay and lesbian website,[26] who commented on the bondage of tradition: "If one were born with shackles, he would think that they were part of him. He would never understand the life and dreams of those who were born without them. He might even fear to lose them." Li is determined to dedicate herself to showing people their shackles and help break them.

She uses this humane approach as one of her ways to urge the government to change. At 4% of the population, the LGBT community is a minority, she points out, "But do we as a society abandon them just because they are a minority? You know that we have more than 50 minority peoples in our country, and the government has done a darn good job keeping them in the fold. It could extend this status to the LGBT community." Li challenges the social tradition and even the moral standard, but she does not—at least directly—challenge the State. Like a good educator, she chooses to nudge the State toward transferring its success with ethnic minorities to this sexual minority group because she is fully aware of the double-edged sword of State power. It could promote her cause as well as harm it.

Unlike a set of legal codes, government policy toward sex and sex studies offers no blueprint either to follow or to challenge. It is thus open to case-specific subjective interpretations. From Li's personal experience, getting funding for sex studies has never been easy. Most of the time, Li has to divert funds for less sensitive projects—studies on traditional family structure, women's roles in villages, etc.—to her sex-related studies. And, once the work is done, the publication of books on sexology also requires some luck. As early as 1998, when *The Subculture of Homosexuality* came out, an order came "from above" to hold up the distribution and destroy all books on site. But the order came too late, and the book ended up reaching the market as planned. One of her latest manuscripts was not as lucky. No one has yet agreed to publish *A Study of Sexual Discourse in New China*, a study of all the articles on sex that appeared in the country's mouthpiece, *The People's Daily*, from 1949 to 2010.

Finding "Self" in Pleasure and Happiness
Li remains committed to offering a different voice. She compares Chinese traditions to those of the West, and seeks inspiration from her philosophical studies. In the nearly one hundred years since the May Fourth Movement of 1919 when youth eloped to pursue love and to control their own lives, self-expression in sexual practices has continued to be an uphill battle. "The State is able to suppress sexual expression in the name of the State and culture," Li reasons, "because the notion of 'self' has not yet fully formed." In Chinese tradition, family is the fundamental unit of human society, whereas in the West, the individual plays the central role. Hence, while both Confucius and John Locke believe that humans are born perfectible (the Lockean "*tabula*

rasa"), they differ in their approach to the individual in society. Confucius advocates "conquering oneself and returning to rituals" (meaning social etiquette), subjecting individual rights to those of the familial and social group. Locke believes individuals are by nature free and equal. They have rights to life, liberty, and property, and even rights to replace a government should the government fail in its social contract to protect individual rights.

Li builds her studies on the common ground between Confucius and Locke: the belief that human beings are born with an improvable nature that can be shaped, enriched, and perfected by experience and education. As a scholar, she does not focus on right or wrong. Sex research and advocacy are about waking up the self and promoting knowledge so people will be able to make their own informed choices and live fuller lives. "Life is so short. I want to fill it with freedom and beauty," Li says, explaining why she continues to try to publish her studies on sexuality.[27]

The philosopher that Li admires most is Michel Foucault. As a scholar and as a social advocate, Li strives to educate the population as she campaigns for the inclusion of social groups marginalized by their sexual orientations. In books and articles, she makes distinctions between sex and gender, individual and society: Sex is a natural endowment and gender a socially assigned role. Li explains the Foucauldian view that homosexual or LGBT practices are explorations of different possibilities of human relationships and different ways of achieving pleasure and happiness.

In the same vein, she introduces Queer Theory in some of her more academic writings, noting the subversive nature of the homosexual lifestyle in the dominant heterosexual society as homosexuals refuse to be compromised and assimilated. They challenge the status quo by taking a stand and fighting for legitimacy and social acceptance of their way of life. In this sense, the conscious choice of such a lifestyle poses more of a threat to society than the sexual preferences themselves, because this choice comes from defiance and the desire to transcend traditionally prescribed relationship norms. Activist homosexuals demand social acceptance of this "invention" (Foucault's word) of lifestyle as an alternative human relationship. Many of these discussions are found in Li's books, such as *Foucault and Sexuality*, *On Sexuality*, and *Queer Theory*.

Regardless of the society or historical time, the issue of sex always brings the individual and society into sharp focus. As Li says, "For as long as there is sex, there is anti-sex." Li's research shows that sexual oppression can come from many sides if one looks across cultures and time: political leaders who are bent on social control, religious leaders who urge and demand sublimation of human desires to a higher spiritual realm, social decorum such as in the Victorian era that regarded sex as a base impulse to be curbed, and Chinese traditions that directly link sex to procreation (a Confucian view) or to the enhancement and lengthening of life (a Daoist approach). Procreation makes sex proper, love makes it good, and pleasure makes it dangerous. And in the first 30 years of Communist China, along with the denial of self, seeking pleasure—indeed, including other personal pleasures beyond sex—acquired a negative connotation. Generations that have come of age in that time have learned to defer their self and pleasure to the greater good. It will take time to disassociate shame from the pursuit of personal pleasure, especially among the older generations or those who prefer to assume a moral high ground.

A Different Kind of Proletarian in the Market Economy

The desire and demand for social acceptance of a lifestyle choice is pertinent to another marginalized social group that Li is fighting for: prostitutes. On an unusually brilliant spring day in Beijing, Li and I meet at Taoran Park in the neighborhood where she grew up. New retirees comparable to her in age hang out in the shade on this weekday afternoon. They are the generation that was born and came of age with the new China. Before their time, prostitution was just the transactional commodification of a natural human need. But when the new Communist China took over the country, prostitution was among one of the first things targeted for eradiction from the new society. It was made illegal by the 1950 Marriage Law.

This was part of the CCP's show of determination to liberate women. In a society deeply entrenched in the belief that chastity is women's ultimate virtue, the eradication of prostitution was a statement as well as an action akin to an act of chivalry that would redeem prostitutes and then burn down the whorehouse. But Li thinks the oldest profession in human history will never be eradicated in China or anywhere else. One of the reasons prostitution was not a notable social phenomenon before the time of Reform and Opening in the first 30 years of the PRC, aside from the fact that it was illegal, was due to an economic reality: "When everyone earns 30–40 yuan a month, who's got spare change for hiring a prostitute?" Yet the human demand is always there, she believes, and now—with the ratio of 120 men to every 100 women in the youngest sexually active group of the population, by some estimates—the demand has increased.[28] To exacerbate the problem, men's stronger social positions and their higher income have made it easier for them to obtain sex with money or coercion.[29]

As in every society, prostitutes enter the profession for different reasons: some to make a living, some for pleasure, and still others because they are forced. While those who are forced into prostitution often draw sympathy from the public, Li noted that this lifestyle as a whole has never had a place in the PRC. The government's criminal law still makes prostitution illegal, women's organizations still refuse to accept the neutral term "sex workers," moralists who prefer to hold on to chastity do not even treat prostitutes as fellow humans, and corrupt law enforcers have law on their side as they levy fines at will. Meanwhile, most prostitutes in China are in the "ten yuan sex shops," where each transaction costs 1.5-3 USD, and the price is often negotiable depending on whether or not condoms are used.[30] These shops primarily cater to the large number of migrant workers who have left their families in poor villages to work for minimum wages in the city. Both the customers and the prostitutes in the ten yuan sex shops make up the lowest rung of society. This is a world away from a much smaller number of high-end "wholesale" prostitutes (as Li characterizes them) whom the rich and powerful keep in their various residences. Right now, the official estimate suggests there are 6 million prostitutes across the country, but Li believes that the number is closer to 20 million, as compared to the often-quoted number of one million in the United States.[31]

The plight of prostitutes prompts Li to reflect on her education in China's revolutionary past: "In the *Communist Manifesto*, Marx declared that what the proletarians lost in the revolution was their bondage, and what they had gained was the whole world," Li says with a faint smile as she notes the irony of applying one of the many Communist mantras to a different era. "If this is the case, prostitutes have to be a different kind of proletarian." Because, she explains, after they lost their bondage the prostitutes of today's China have entered a cage filled not only with the constant threat of sexual diseases such as AIDS, but also one that attracts extortion and abuse, and one that is deprived of basic human dignity and recognition that is worthy of legal protection. "Thirty-eight percent of the unsolved homicide cases in Beijing involve prostitutes," Li quotes from memory of her own recent data.[32] Having drifted far from their social network due to shame, these prostitutes have basically become social outcasts and are thus easy targets for predators.[33]

When asked whether she still hopes to legalize prostitution, she says this mission would currently be "extremely difficult." She mentioned a lawyer named Chi Susheng who was also an NPC delegate. Chi made several attempts to propose the legalization of prostitution, but they fell on deaf ears. "Remember," Li explains, "abolishing prostitution was one of CCP's landmark achievements for the new China. The government will never go back on its own words." With this realization, Li has since modified her demand from legalization to decriminalization of prostitution. Ultimately, she reasons, the goal is to solve the problem in real life rather than just make an academic statement. Referring to the prostitutes, she says: "Clearly, to fine them and to send them to reform camps is not a solution. They have to make a living. The solution is to give them an employable skill. But right now, you have to accord them basic human dignity and not consign them to an unlawful existence."[34]

Prostitution, Li points out, is a result of inequality between men and women and between haves and have-nots.[35] But when the majority of women themselves value chastity over life, the battle goes beyond fighting male dominance in society. Li and her followers have to convince women that the idea of women's chastity as a badge of honor is a product of a male-dominated society and that every adult has the right to decide how to use his or her own body: for example, everyone can use sex to achieve pleasure or to make a living (as prostitution enables them to do). This insight empowers women to establish a self equal to men's. But this is a new perspective for many women, especially for prostitutes. To give prostitutes the freedom to do what they choose, or end up doing, is the first step to giving them back their dignity.[36]

"Sex is the Key to Understanding the Heart of a Society"

A sexual revolution is sweeping across the country in a way that people here have never seen before. Using the Internet as a public forum, people are sorting out their confusion, doubts, and comfort zones in their private lives anonymously, immediately, and often without reservation. The mainstream media has also started to pick up the taboo subject of sex. Li is keeping up her mission of bringing attention to a different voice while getting her message into the mainstream media to influence the public debate. Her 48 books published to date include translations of Western classics written by philosophers from Aristotle to Foucault; studies, observations, and comments on sexuality, women, family, and more recently, the discrepancy between the rich and the poor.[37] Her online blog contains 59 pages so far[38] of observations, comments, and sometimes strongly-worded opinions. She updates her blog every couple of days, reflecting on life in terms not limited to the realm of sex or love. As she increasingly becomes an icon of the sexual revolution in China, she enhances that reputation by holding an online salon every Friday to discuss and answer questions on sex, sexual relationships, and love.[39] It has not been easy to recruit experts to talk openly about sex issues, she confesses, so she has only a limited group of scholars who are willing to participate. As a familiar name on LGBT sites as well as in mainstream media both inside and outside of China, Li seems to enjoy her free citizen status in her retirement.

Li's prolific writing and social activism are noted both at home and abroad. One of the latest of her profiles appeared on Radio Netherlands Worldwide (May 6, 2013), titled "Chinese Sexologist Sparks Debate on Prostitution."[40] The debate was over a blog Li posted reviewing the British TV series *Secret Diary of a Call Girl*.[41] The point in contention was Li's comment that legalized prostitution, as seen in Britain, offers prostitutes a better working environment and is a better legislative solution. For Li, Western tradition, scholarship, and the women's rights movement have much to offer, both in modernizing Chinese customs and in preparing for the battles ahead. "Human nature is more or less the same while customs differ," she reasons. As she casts light on set ways of governing and thinking, particularly with regard to sexuality in China, she unabashedly holds up the West as the source of her insight.

After returning to Li's villa from our walk, I ask her about her being called a "fighter." Li dismisses the term in her gentle voice with a smile: "I don't shed blood, I just do my research. Moreover, I simply speak out about what has already happened in real life and try to defend what I believe human beings are entitled to do." She takes a deep breath and starts again with a sigh of resignation: "I can turn away from the sufferings of others and enjoy my own peaceful life. But when I see people's basic rights get trampled, I become indignant. The idealism that is part of the DNA of our generation pushes me to speak out to make our society a better one. But," she pauses as if collecting her thoughts, "an inevitable goal of sex studies is to change misguided and incorrect concepts and to challenge society." Similarly, Li is dismissive of the term "avant garde" when it is used to refer to her views and advocacy: "If I had said in 1900 that we shouldn't bind women's feet, I would've been 'ahead of my time,' and if I had said at the time that men and women could kiss before they married, I would also have been 'ahead of my time.' If we wait for an appropriate time to make changes, then there is really no hope for social progress."

The drizzle has stopped as she rummages through her piles of author's copies to look for hard-to-find books. She offers to give me a few. On the wall is a framed photo of her son, Zhuang-zhuang. Li and her husband decided not to have children in order to devote more time to the work they loved. But years after Wang's untimely death, Li adopted a son from an orphanage. She named him Zhuangzhuang, meaning "strong, strong." She hopes Zhuangzhuang, now 12, will grow up to enjoy a more open society with fewer and less stringent restrictions on personal expression. From the wall, Zhuangzhuang's radiant and innocent smile seems to warm the room.

The world Li is hoping to deliver to Zhuangzhuang is rapidly evolving. Coming out of the long Confucian tradition and the PRC's first 30 years of revolutionary rhetoric, sex and individuality have finally emerged as topics of free online discussion. "Sex is the key to understanding the heart of a society," Li explains, and she has made it a branch of sociological study in China.

As an American-trained sociologist, Li had not a moment of hesitation about returning to her native country upon the completion of her doctorate at the University of Pittsburgh. "China is my society," she says. In this ancient society, she has made her mark. She is the first female sexologist in China's long history. She openly talks about sex as a scholar and educator, chipping away at the taboo encrusted upon the subject. Using Western learning and Western philosophy, she excavates for gems and exposes detritus in her own cultural heritage to show her own people the universality of human nature.

After some bumpy moments caused by the initial shock of Li's introduction of such controversial topics, the Chinese government and society have been generally accepting of her. She retired as a professor from the Academy of Social Sciences and is respected by most people in society for her knowledge, courage, and vision. Although none of her students have selected Li's most controversial topics as their major field of study,[42] she nevertheless puts her faith in an increasingly open society where sex is no longer taboo and personal pleasure is a justifiable pursuit.

With only four sexologists, including Li,[43] China has

managed to come a long way in its sexual revolution. Many of the issues Li has pushed into the mainstream debate in China are still controversial even in the United States. Prostitution is illegal in most states, pornography's boundaries are in flux, and, of course, same-sex marriage has also been an ongoing debate in and out of the Supreme Court.[44] Yet she may have reason to be cautiously optimistic. During the last 30 years of the country's economic boom, Chinese criminal law, which started at the same time as the boom, has undergone eight revisions in an effort to keep pace with rapid social change, and the media has turned from hostility to greater sympathy on these issues. The friendlier atmosphere created by the media's civil tone allows more meaningful discourse to take place in public as well as academic circles. "Being able to say what I say and publish what I've published shows how far the country has come along," Li says. She is a gentle and scholarly rebel. Her goal, as she insists, is not to tear down the government, but to build it up to lead to a more open society where individuals have more room to define and exercise their inalienable rights.

Chapter 2 Notes

1. Li Yinhe is in the section "People to Know: 50 Movers and Shakers in Today's China," *Asiaweek* (September 24, 1999), 74. This being the year of 50th anniversary of the founding of the People's Republic of China, many names on the same list have indeed become history makers and remain well known today: Wen Jiabao (premier 2003–2013); Wu Jinglian (economist); Tung Chee-hwa (the first chief executive of Hong Kong 1997–2005), and artists such as Zhang Yimou and Gong Li.

2. Li Yinhe posted this letter on her blog on March 9, 2012 (see http://blog.sina.com.cn/liyinhe).

3. Before the first revision of the criminal law in 1997, there was an item called "hooliganism" that carried a sentence of three years to death. In her *On Sexuality* 性的 (Chinese Youth Press 中国青年出版社, 2009), Li Yinhe presents many cases with different degrees of sentencing.

4. In the updated civil law called The Public Security Administration Punishment Law of the PRC 中华人民共和国治安管理处法 effective 2006, items 66–69 deal with sex-related activities that include prostitution and distribution of sex-related publications. The notable part is in item 69.3: the punishment for group sex, part of the crime of "hooliganism" in the 1979

Criminal Law, is "detention between 10–15 days, and fines ranging from 500 to 1000 yuan" (http://www.law-lib.com/law/law_view.asp?id=97597).

5.   Please consult the bibliography at the end of this book for a list of Li Yinhe's publications.

6.   The Chinese title of this book is *Loving You is Like Loving Life* 你就像生命 (Zhaohua Press 朝华出版社, 2004). This same book was republished under a different title: *Fall in Love If You Wish* 假如你愿意你就恋吧 (Shanxi Normal University 西师范大学, 2006).

7.   The Chinese title for the book is 他的世界 – 中国男同性恋群落透视 (Hong Kong Cosmos Press 香港天地公司, 1992).

8.   This is a recurring topic in many of her books, articles and speeches. Some of her books that expound upon this topic include: *Sexuality and Marriage of the Chinese* 中国人的性与婚姻 (Henan People's Press 河南人民出版社, 1991); *Procreation and Chinese Village Culture* 生育与中国村落文化 (Chinese Academy of Social Sciences 中国社会科学出版社, 1994); *Transformation of Marriage and Family in China* 中国婚姻家庭及其变迁 (Heilongjiang People's Press 黑龙江人民出版社, 1995).

9.   Aside from her first work on homosexuality in 1992, *Their World*, she has also published many studies and translations affirming equal rights for homosexuals. Many of these efforts were intended to explain and educate the general public on human sexuality, particularly homosexuality. Her works on this topic are wide ranging: there are direct translations from major Western publications such as *An Introduction to Major Western Sexual Studies* 西方性学名著提要 (Jiangxi People's Press 江西人民出版社, 2002), and her delineation and interpretation of works by Western scholars: *Queer Theory: The Sexual Trend in the West in the 1990s* 酷儿理 – 西方90年代性思潮 (Shishi Chubanshe时事出版社, 2000), *Foucault and Sexuality – Reading Foucault's "The History of Sexuality"* 解福柯与性 – 解福柯'性史,' (Shandong People's Press 山东人民出版社, 2001); as well as the results of her own research on sex and homosexuality: *The Subculture of Homosexuality* 同性恋亚文化 (China Today Press今日中国出版社, 1998); *On Sexuality* 性的 (Chinese Youth Press中国青年出版社, 1999); *Sex/Marriage— East and West* 性／婚姻 — 东方与西方 (Shaanxi Normal University Press西师范大学出版社, 1999).

10. The results of the 2008 poll can be found on Li Yinhe's blog (June 17, 2008 http://blog.sina.com.cn/s/blog_473d533601009vfr.html). The same issue is also discussed in *Li Yinhe's Reflections on Sexual Studies* 李银河性心得 (Shidai Wenyi Press 时代文艺出版社, 2008).

In an article on the Gallup website by Frank Newport, "Six Out of Ten Americans Say Homosexual Relations Should Be Recognized as Legal," 56% of Americans thought that homosexual men and women should have equal rights when Gallup started its survey in 1977, and "as recently as 1992, fewer than four in five Americans felt homosexuals should be given equal treatment in hiring." As for whether or not homosexuality should be an acceptable lifestyle, the same article pointed out that in 1982, 34% American said yes. See: http://www.gallup.com/poll/8413/six-americans-say-homosexual-relations-should-recognized-legal.aspx. Also see another Gallup site posting statistics from their polls at: http://www.gallup.com/poll/1651/gay-lesbian-rights.aspx

11. The Gallup site shows this set of numbers: in 1996, 27% said the marriage should be valid, and 68% against; in 2008, that number moved to 40% for and 56% against (http://www.gallup.com/poll/1651/gay-lesbian-rights.aspx).

12. The "half of the sky" phrase comes from Mao Zedong in the early years of the Cultural Revolution (1966–1976). His complete saying is: "Women can hold up half of the sky."

13. Deng Yingchao's intervention has been widely circulated. But due to the sensitivity of the subject matter and lack of media openness at the time, news reports cannot be found. What was reported was that Deng Yingchao was one of the presenters at the ceremony for the "Outstanding Young Creative Actors Award" issued by the Ministry of Culture for the year 1980, and Chi Zhiqiang was one of the receivers of this award. See "A Must-Know Acting Career: The Young National Star Actor Chi Zhiqiang 于全国优秀青年男演志强，你不得不知的演艺经历" (WWW.72177.com [accessed April 9 2015] and http://news.72177.com/a/201509/042070640.shtml [accessed March 28, 2016]).

Chi Zhiqiang's return to the stage after serving his prison time refreshed the public memory of his dramatic fall in 1984. Many online sites today recount and repeat the same story. One example can be found at http://news.qq.com/a/20080228/003732.htm.

14. See the article in *Chengdu Commercial Newspaper* 四川新网－成都商报 on January 2, 2009: "Chi Zhiqiang: If I Were Born 20 Years Later, I Certainly Would Not Have Had to Go to Jail 志强：如果晚生二十年，我一定不会坐牢." Sina. January 2, 2009 (see http://ent.sina.com.cn/m/c/2009-01-02/10432325355.shtml [accessed March 28, 2016]).

15. The crime of "hooliganism" was #160 in the PRC's first Criminal Law 刑法 drafted in 1979, effective January 1, 1980. Serious cases carried prison terms of seven years or more. Under Article 160, it says: "Where an assembled crowd engages in affrays, creates disturbances, humiliates women or engages in other hooligan activities that undermine public order, if the circumstances are flagrant, the offenders shall be sentenced to fixed-term imprisonment of not more than seven years, criminal detention or public surveillance. Ringleaders of hooligan groups shall be sentenced to fixed-term imprisonment of not less than seven years" (see http://www.opbw.org/nat_imp/leg_reg/China/CRIMINAL_LAW.pdf).

16. This 1997 version of the Criminal Law stipulates in article 301 that: "Whoever takes a lead in assembling a crowd to engage in promiscuous activities or repeatedly participates in such activities is to be sentenced to not more than five years of fixed-term imprisonment, criminal detention, or control" (see http://www.lawinfochina.com/display.aspx?lib=law&id=354&CGid).

17. The Chinese title is: 中华人民共和国治安管理处条例 (1987); 中华人民共和国治安管理处法 (2006). Among the three offenses under article 69 that carry a punishment of "detention between 10–15 days, along

with 500–1,000 yuan fine" is this provision: "Participating in assembling a crowd to engage in promiscuous activities" (see http://www.cecc.gov/resources/legal-provisions/public-security-administration-punishment-law-chinese-text). The 1987 version, "Public Security Administration Punishment Regulations of the People's Republic of China 中华人民共和国治安管理处条例" can be found at http://baike.baidu.com/view/33749.htm (accessed March 2016). This same law was further amended in 2012, with no change to the sentencing of sex-related offenses: http://www.gov.cn/flfg/2012-10/26/content_2253934.htm (accessed March 2016).

18. See *Diagnostic Standards of Psychiatric Diseases*中国精神疾病断准, available online at http://www.wendangwang.com/doc/46ff7fb9392d-3c625b997fc3/2. In the years leading up to this milestone, Li Yinhe had been the most outspoken scholar arguing that homosexuality was a lifestyle issue and not a disease. Starting from her first book, *Their World* (1992), devoted to homosexuality, Li Yinhe appealed to the public with real stories from her interviews that showed the hardship, the internal torment, and the social and familial pressures to live a "normal" heterosexual life. She framed the issue as one of an emotional and physical desire that cried out to be fulfilled—something an average person could easily identify with. She expounded on the issue of homosexuality with many other publications in the intervening years: *Sociology of Sex: Human Sexual Behavior* 性社会学 – 人类性行, 1994; *The Subculture of Homosexuality* 同性恋亚文化, 1998; *The Subculture of Sadomasochism*虐恋亚文化, 1998; *Queer Theory: The Ideological Trend in Sexology in the 1990s West* 酷儿理——西方90年代性思潮, 2000; *Foucault and Sexuality: Reading Foucault's "History of Sexuality"* 福柯与性——解福柯性史, 2001. In both her writing and public speech, she also made a case for the universality of homosexuality, pointed out that it could be found in societies across time and cultures despite their relatively small number in any given society.

19. Versions of the ordinance were passed in several cities in the United States but struck down by courts, who found it to violate freedom of speech protections in the First Amendment to the United States Constitution. Dworkin and MacKinnon lay out their case in their co-authored book *Pornography and Civil Rights: A New Day for Women's Equality* (1988), available online at http://www.nostatusquo.com/ACLU/dworkin/other/ordinance/newday/TOC.htm.

20. The same discussion and data can be found in her book *On Sexuality*, chapter 3.

21. A TV clip of this story about Tu Shiyou was posted on this popular website, available when I first accessed it in early 2013: http://tv.people.com.cn/GB/150716/156088/156089/17089344.html. But as the sensation subsided, the clip was cycled out of the news.

22. These numbers can also be found in chapter 6 of *Sexuality and Gender* (2005).

23. This number can be found, among other places, in Li Yinhe's blog on July 13, 2010: http://blog.sina.com.cn/s/blog_473d53360100k3ou.html. The

blog has a long discussion on the topic of chastity inspired by her walk on a trail in a town called "Chastity Tablet Village" in Anhui province.

24. This article can still be found at http://news. qq.com/a/20060722/000730.htm. Li Yinhe responded in an interview with *Youth Weekly* to clarify her views: http://culture.people.com.cn/GB/22219/4707058.html.

25. See Li Yinhe's August 6, 2006 blog: http://blog.sina.com.cn/s/blog_473d533601000592.html.

26. For the link to the site, please see http://www.danlan.org/.

27. Li Yinhe also wrote a blog titled "Why do I want to study sexuality?" in the early days of her blogging (January 8, 2006): http://blog.sina.com.cn/s/blog_473d5336010001o7.html. In the blog she says that sexuality is part of our everyday life and yet it is absent in open discourse. This leaves room for manipulation by those in power in a male-dominant society. She believes that many aspects of human sexuality are related to basic human rights, and that they deserve serious scholarly study.

28. Statistics from the US Central Intelligence Library show the following numbers for China in 2015:

at birth: 1.15 males/female

0-14 years: 1.17 males/female

15-24 years: 1.13 males/female

25-54 years: 1.04 males/female

55-64 years: 1.02 males/female

65 years and older: 0.92 males/female

total population: 1.06 males/females (2015 est.)

(Last retrieved on March 28, 2016, at https://www.cia.gov/library/publications/the-world-factbook/fields/2018.html).

Also, according to a 2015 news release issued by Chinese National Health and Family Planning Commission 中华人民共和国国家生和划生育委会 , "National Gender Ratio at Birth Shows a 'Six-year Consecutive Decline' 我国出生人口性别比出'六降'," the gender ratio at birth in the years 2008–2014 are as follows:

2008:120.56 males/female,

2009: 119.45 males/female,

2010: 117.94 males/female,

2011:117.78 males/female,

2012: 117.70 males/female,

2013: 117.60 males/female,

2014: 115.88 males/female

(Last retrieved November 2015, at http://www.nhfpc.gov.cn/jtfzs/s3578/201502/ab0ea18da9c34d7789b5957464da51c3.shtml).

29. Li Yinhe's book *Sexuality and Gender* (2005) is a comprehensive study of gender imbalance across societies and over time.

30. A good report on the Ten Yuan Sex Shops 十元店 can be found online at http://news.163.com/12/0427/18/80490KUP00011229_all.html. The title of this Chinese article is "Investigation on Ten Yuan Shop Sex

Workers: Women Farmers View Receiving Guests as Working in the Fields 十元店性工作者查：农视接客种田" (April 27, 2012 [accessed March 20, 2016]).

31. Keeping these "wholesale prostitutes" has become such a prevalent social phenomenon that it became a target of attack and ridicule. The young, avid social blogger Han Han, for example, has a much quoted satirical comment: ". . . it is cheap to go for a $100 prostitute, but to spend one million on a show girl is elegant . . ." It has become one of Han's best known quotes, and can easily be found in places such as *Wikiquote*: http://en.wikiquote.org/wiki/Han_Han.

32. In her book *On Gender* (2007), Li Yinhe quotes a 2004 figure of 57% (p. 228).

33. Li alludes to the well known activist nicknamed "Hooligan Yan 流氓燕," whose real name is Ye Haiyan 叶海燕. In order to bring attention to the plight of prostitutes, she worked as a prostitute for three days and publicized what she learned in the low-end brothel. Hooligan Yan's stories and her activism can easily be found online. She also has a few blog sites including: http://blog.ifeng.com/1403777.html.

34. The transcript of a more recent interview on the topic in 2014 appears on Ifeng.com: http://news.ifeng.com/exclusive/scholar/detail_2014_02/11/33683553_0.shtml. In the interview, she calls for decriminalization of prostitution. A transcript of an interview on the same subject with the lawyer Chi Susheng on March 13, 2012, can be found at http://www.360doc.com/content/12/0313/19/6795100_194086055.shtml.

35. Li has a long blog on this subject on February 16, 2011, also in Chapter 6 of *On Gender* (2007), among many other places.

36. In Chapter 4 of *On Sexuality* (2003), Li devoted two sections to women's rights and their own sexuality: "On Prostitution" and "Women's Rights and Sexual Politics." In these two sections, she quoted widely from Western sources: A. Soble, *The Philosophy of Sex*; P. R. Abramson and S. D. Pinkerton, *With Pleasure*; L. Lemoncheck, *Loose Women, Lecherous Men: A Feminist Philosophy of Sex*; A. Assiter and A. Carol (ed.), *Bad Girls and Dirty Pictures*; L. Bell (ed.), *Good Girls/Bad Girls: Sex Trade Workers and Feminists Face to Face*; L. Coneney et al. *The Sexuality Papers: Male Sexuality and the Social Control of Women*; D. C. Stanton (ed.) *Discourses of Sexuality: From Aristotle to AIDS*; D. Cooper, *Power in Struggle: Feminism, Sexuality and the State*; David M. Halperin, *Saint Foucault: Towards a Gay Hagiography*; and M. Foucault, *Philosophy, Culture, Interviews and Other Writings*, etc.

Li Yinhe herself also has a book devoted to the study of women's rights: *The Rising Power of the Chinese Women* 中国女性利的崛起 (Chinese Social Science Press; 1997).

37. Li Yinhe provided me with an updated list of her books in spring of 2015. Since then, her own autobiography has been published online, making it 48. Books reprinted under different names are indicated as such. On average, Li has kept up with the rate of at least one publication per year.

The main difficulty in determining the exact number of her publications is simply their availability—or more often the lack thereof—through regular channels. Regular bookstores usually don't carry them, and online sources are hit or miss. Moreover, because of the long time span of her publications, many have limited copies and are out of print soon after publication. By her own assessment, the library that is most likely to have a relatively complete collection of her works would be the Peking University library.

Here is the list Li Yinhe made available to me:

1. *A Brief History of the May Fourth Movement* 五四运动史, (co-author), 山西人民出版社 1978.

2. *Methodology in Sociology Studies* 社会研究方法, (translated from Babbie, Earl. *The Practice of Social Research*. Wadsworth Publishing Company. Belmont, California, 1983), 四川人民出版社 1987.

3. *Introduction to Modern Sociology* 代社会学入门, (translator), 中国社会科学出版社 1987.

4. *Sexuality and Marriage of the Chinese* 中国人的性与婚姻, Henan People's Press 河南人民出版社, 1991.

5. *Their World: A Look into Chinese Male Homosexual Community* 他的世界－中国男同性恋群落透视, Hong Kong Cosmos Press 香港天地公司, 1992; Shanxi People's Press 山西人民出版社, 1993.

6. *Procreation and Chinese Village Culture* 生育与中国村落文化，Oxford University Press, Hong Kong. 1993；Chinese Academy of Social Sciences 中国社会科学出版社，1994.

7. *Sociology of Sexuality: Human Sexual Behavior* 性社会学－人类性行, (co-author), Henan People's Press 河南人民出版社, 1994.

8. *Transformation of Marriage and Family in China* 中国婚姻家庭及其变迁, Heilongjiang People's Press 黑龙江人民出版社, 1995.

9. *Sexuality and Love of Chinese Women* 中国女性的性与, Oxford University Press, Hong Kong, 1996.

10. *Women: The Longest Revolution: Selected Readings on Contemporary Women's Rights Literature* 女－最漫长的革命，当代西方女主义理精选，(editor) Joint Press 三联店, 1996; Chinese Women's Press 中国女出版社, 2007.

11. *The Rising Power of the Chinese Women* 中国女性利的崛起, Chinese Social Science Press 中国社会科学出版社, 1997.

12. *The Subculture of Homosexuality* 同性恋亚文化 Today's China Press 今日中国出版社, 1998.

13. *The Subculture of Sadomasochism* 虐恋亚文化. Today's China Press 今日中国出版社, 1998.

14. *The Chinese Women's Emotions and Sex* 中国女性的感情与性, Today's China Press 今日中国出版社, 1998.

15. *On Sexuality* 性的, Chinese Youth Press 中国青年出版社, 1999, 2003.

16. *Sex/Marriage: East and West* 性／婚姻－东方与西方, Shaanxi Normal University Press 西师范大学出版社, 1999.

17. *The Debate on the Revision of Marriage Law* 婚姻法修改争, (editor), Guangming Daily Press 光明日报出版社, 1999.

18. *Queer Theory: the Ideological Trend in Sexology in the 1990s West* 酷儿理论 – 西方90年代性思潮, Shishi Press 时事出版社, 2000.

19. *Enjoy Life* 享受人生, Baihuazhou Arts Press 白花洲文艺出版社, 2000.

20. *Farmer Migration and Gender* 农民流动与性别, (co-author) Zhongyuan Farmer's Press 中原农民出版社, 2000.

21. *Grandfather and Grandson: Case Studies of Chinese Families* 一之 – 中国家庭系个案研究. (co-author), Shanghai Cultural Press 上海文化出版社, 2001.

22. *Foucault and Sexuality: Reading Foucault's "The History of Sexuality"* 福柯与性 – 解福柯'性史,'" Shandong People's Press 山东人民出版社, 2001.

23. *Abstracts of Western Classics on Sexology* 西方性学名著提要, (editor) Jiangxi People's Press 江西人民出版社, 2002.

24. *On Feminism* 女性主义. Wunan Press 五南出版公司, Taiwan, 2003.

25. *Report on the Study of Sexual Culture* 性文化研究报告, Jiangsu People's Press 江苏人民出版社, 2003.

26. *Sexual Love and Marriage* 性与婚姻, Wunan Press 五南出版公司, Taiwan, 2003.

27. *Female Emotions and Sexuality* 女性的感情与性, Wunan Press 五南出版公司, Taiwan, 2003.

28. *On Gender Relations* 两性系. Wunan Press 五南出版公司, Taiwan, 2004; Huadong Normal University Press 华东师大出版社, 2005.

29. *The Poor and the Rich: the Diversification of Chinese Urban Families* 人与富人——中国城市家庭的富分化, Huadong Normal Uinversity Press 华东师范大学出版社, 2004.

30. *Forum on Feminism* 女性主义坛, Proceedings of the conference 会文集, 2004.

31. *Thinkers' Talk: A Collection of Wang Xiaobo and Li Yinhe* 思想者 —— 王小波李银河双人集, Cultural and Arts Press 文化艺术出版社, 2005.

32. *On Feminism* 女性主义, Shandong People's Press 山东人民出版社, 2005.

33. *You Are So In Need of Comfort: A Dialogue on Love* 你如此需要安慰 —— 于的对, Contemporary World Press 当代世界出版社, 2005.

34. *Loving You is Like Loving Life* 你就像生命, Zhaohua Press 朝华出版社, 2004; Republished under a different title: *Fall in Love if You Wish* 假如你愿意你就恋吧, Shanxi Normal University 西师范大学, 2006. (Li does not regard this one as her book, as it contains mostly love letters from her husband Wang Xiaobo).

35. *Li Yinhe's Self-Selected Works* 李银河自选集, Inner Mongolia University Press 内蒙古大学出版社, 2006.

36. *On Gender* 性别, Qingdao Press 青出版社, 2007.

37. *Qixi, Folk Customs, and the Culture of Emotions* 七夕 / 民俗 / 情感文化, (editor), China Radio/TV Press 中国广播电视出版社, 2007.

38. *Live with Grace and Gentleness* 以温柔优雅的度生活, Chinese Women's Press 中国女出版社, 2007.

39. *20 Lectures on Sexual Love* 性20, Tianjin People's Press, 2008.

40. *Li Yinhe's Reflections on Sexual Studies* 李银河性学心得, Times Arts

Press 时代文艺出版社, 2008.

41. *The Essence of Sociology*社会学精要, Inner Mongolia Press内蒙古出版社, 2009

42. *The Women of Houcun: Gender and Power in the Countryside*后村的女人 – 农村性别利系, Inner Mongolia Press, 2009

43. *Introduction to Sexology*性学入门, Shanghai Social Sciences Press 上海社会科学出版社, 2014.

44. *My Social Observation*我的社会察, Chinese Labor and Commerce Union Press 中华工商联合出版社, 2014.

45. *A Study of the Language on Sexuality in the New China*新中国性研究, Shanghai Social Science Press上海社会科学出版社, 2014.

46. *My Life's Philosophy*我的生命哲学, Chinese Labor and Commerce Union Press中华工商联合出版社, 2014.

47. *Readings for My Soul*我的心灵, Chinese Labor and Commerce Union Press中华工商联合出版社, 2014.

38. As of March 2015, the total number of her blog pages is 74. Her blog site is: http://blog.sina.com.cn/liyinhe.

39. The link is: http://z.t.qq.com/sexology/salon04.htm. The sessions usually follow this formula: Li Yinhe hosts the session with a special guest for the day. Netizens pose questions to the guest, the special guest responds, and Li Yinhe adds her comments from time to time.

40. The link is: http://www.rnw.org/archive chinese-sexologist-sparks-debate-prostitution. Since the completion of this article, *The New York Times* has also published a profile by Andrew Jacobs on March 16, 2015: "Sex Expert's Secret is Out, and China's Open to It."

41. *Secret Diary of a Call Girl* is a British television drama that many critics compare to the US television show *Sex in the City*. Set in London, the British show revolves around a woman named Hannah Baxter. Hannah is a college graduate living a seemingly normal life (although she seems to have difficulty in holding down a job) yet at the same time leading a secret life as a high-end call girl using a pseudonym, Belle. Some feminists in the US criticize the show as objectifying women, others accuse it of glamorizing and misrepresenting prostitution. Li Yinhe's blog on the show was posted on June 28, 2013: http://blog.sina.com.cn/s/blog_473d53360102f1m4.html.

42. Li has mentioned that when her students did propose more sensitive topics, they couldn't get them approved, and so had to switch to other less sensitive ones.

43. According to Li, the others are: Peng Xiaohui 彭晓, Fang Gang 方钢, and Pan Suiming 潘.

44. Since the completion of this chapter, the US Supreme Court has passed the same-sex marriage law of June 26, 2015.

CHAPTER 3

The Sky Has No Limits:
Wu Zhendong, General Aviation Pioneer

Americans don't give much thought to the fact that they can fly the "friendly skies" with ease because civilians control airspace in the United States. Privately-owned aircraft, corporate jets, and planes flown by companies providing flight-for-hire services can all access US airways without having to ask the permission of military authorities, and without having to negotiate endless amounts of red tape. And most American airports are more than willing to receive these planes.

In China, however, this is not the case. The military is the guardian of Chinese airspace, and has been deeply reluctant to cede that control.[1] The military's argument has always been that "aviation activities can proceed only to the extent that they do not imperil national security, public safety, or even personal safety."[2] Especially with national security said to be at stake, the military has been painfully slow in loosening its grip. Yet when China's economy began accelerating dramatically in the early 1990s, so too did demand for all manner of civilian aviation. With that, of course, came competition for airspace.

General Aviation (GA) refers to the various forms of civilian aviation that do not involve regularly scheduled commercial passenger transport. The GA sector includes activities ranging from private and corporate flights, to airborne search and rescue, air ambulance services, instructional flying, crop dusting, and myriad other industrial applications such as helicopter services for power line inspection, aviation services in the oil exploration industry, and airborne mapping services.

In China, rising demand for general aviation and the military's reluctance to cede partial control over airspace put the government in a bind. It could not risk compromising national security, yet neither could it ignore the increasingly important role that GA would play in a rapidly expanding economy. Simultaneously keeping the peace between these contending parties while keeping the throttle wide open on economic development meant that the government continually had to serve as arbiter over who could access the nation's skies.

When Wu Zhendong registered his aviation service company, Avion Pacific Limited, in 1993, China's GA sector was in its earliest infancy. "Nobody knew what general aviation was back then," he said. "The only activities were agricultural spraying, using Antonov 5's, and some helicopter support for offshore oil platforms. In the latter case, the helicopters were crewed mostly by Westerners partnered with Chinese from state-owned firms like China Southern Airlines, China Eastern Airlines, or the China Aviation Industry Corporation (AVIC). Private ownership was unheard of." Avion Pacific was, and still is, in large part an aviation brokerage firm, guiding clients through the purchase and maintenance of airplanes and helicopters. Setting up this kind of GA firm in the early 1990s was akin to building a castle in the sand. There was no social infrastructure in place to support the firm's development. Similarly, China still lacked a physical infrastructure to support GA with airports that could accept civilian aircraft and maintenance facilities that could service them. And, of course, though the economy was starting to grow rapidly, purchasing power was still limited, so demand for aviation services remained unpredictable. On top of everything else, the government had yet to provide policies and laws that could effectively regulate emerging civilian, non-airline flight operations. Indeed, Zhendong had set up Avion even before GA had its own separate regulatory status in China.[3]

Nonetheless, looking to America and other developed nations, Zhendong felt confident that an advanced economy of the kind China was aspiring to build would eventually demand a strong general aviation sector.[4] He was willing to bide his time, start slowly, and first do business with the government and government-affiliated enterprises. Twenty years later, Zhendong is now poised to expand his business, capitalizing on the government's increasing attention to, and support for, GA. As a "pioneer of Chinese general aviation"—the moniker bestowed on Zhendong by the director of the Beihang University's GA Industry Research Center—Zhendong is today championing GA's expanding role in the new economy.

In the early 1980s, Zhendong was my college classmate in the English department at Zhongshan University. Avion's chief operating officer (COO), Zhuang Bei, was also in the same class. For years, news of their aviation business came to me in bits and pieces from a loose circle of classmates now dispersed all over the world. Learning of my interest in their story, Zhendong and Bei invited me to spend some time in Avion's Shenzhen headquarters. So, in 2012, on the first weekend after China's biggest annual holiday, the Spring Festival, I sat in on a meeting held in Avion Pacific's conference room.[5] The long desk that dominated the room featured three aircraft models along the centerline: a Sikorsky S92, a Sikorsky S76, and a Beechcraft King Air 350i. The first order of business dealt with international delivery of helicopters. Zhendong and Bei were interviewing a highly experienced expatriate helicopter pilot and safety auditor whom they wanted to join King's Aviation, the subsidiary of Avion Pacific set up two years earlier. In addition, they wanted the pilot to arrange plans to deliver helicopters, never an easy thing in China. Sitting across the table from Zhendong and the pilot was Zhuang Bei.

In this particular case, Avion served as a broker for a client's purchase of two S92 helicopters, Sikorsky's most advanced civilian model equipped for offshore operations, with a sales tag of about US $32 million.[6] Avion since 1997 has been the sole sales agent in China for Sikorsky civilian aircraft. The two helicopters needed to be ferried to China from Sikorsky's manufacturing facilities in Stratford, Connecticut. "It's about 7,000 nautical miles," Zhendong says to the pilot interviewee. "Are you able to draft a ferrying plan and find me a couple of experienced pilots?" Ferrying a helicopter from America to China requires not just a complex international flight plan, but also pilots who are certified according to international standards. "And we might have to get it done before the typhoon season in May," Bei says, reminding everyone of the importance of timing. "The helicopters should be in place for weather-related evacuation of offshore oil platforms during the typhoon season. Can the pilots do the ferry operation in April?" After a few more rounds of payment negotiations, they move on to the next item on the day's agenda.

Zhendong is brokering a deal between three parties. Malaysia's national oil company, Petronas, which is collaborating with the Chinese oil company CNOOC on an offshore oil project, needs to find a third party to lease a couple helicopters for offshore operations. Zhendong knows that China Southern Airlines has helicopters available. "They can provide the S76C++," he notes, and wants to know the pilot interviewee's availability to operate them. After assurances that the transaction will be publicly tendered, he establishes a March time frame for the contract.

"Dry or wet?" Bei asks a follow-up question.

"Wet," the pilot replies.

A "wet lease" includes crew, maintenance, and insurance, whereas a "dry lease" involves only the aircraft itself. Foreign companies engaged in resource extraction projects on Chinese territory (i.e., oil and gas exploration, mining, etc.) are required to work with Chinese partners. When Zhendong began Avion Pacific in the early 1990s, he took advantage of this mandatory rule and started brokering aircraft deals for such partnerships. But at that time, he had to lease the helicopters and pilots who could operate them from the US, since Chinese civilian airlines did not offer GA aircraft leasing services. Zhendong found a middle ground, which he jokingly refers to as a "damp lease." This practice involves taking a Chinese pilot on board a wet-leased plane to acquire necessary skills and on-the-job training from the Western pilot who comes with the leased aircraft.[7] The "damp lease" has since been phased out as the leasing practice in general has become more standardized.

Having finished negotiations for the contract work, Zhendong turns to the bigger goal of convincing the interviewee to become an actual employee of King's Aviation, thus fulfilling multiple needs for the company, including pilot, auditor, and trainer. "Wouldn't you like to be the auditor for China's general aviation through King's Aviation?" he asks, pointing out the significance of working with King's as China's skies open to GA.[8] The pilot is inclined to consider the possibility.

The founding of King's Aviation in 2010 represented an important step forward in building Zhendong's GA empire. While Avion started as an aviation consulting company in 1993, it later expanded to aircraft sales, operational and technical support for fixed wing and rotary aircraft (for both onshore and offshore operations), crew training, and project support. King's Aviation, on the other hand, is an aircraft operating company certified by the Civil Aviation Administration of China (CAAC). In laypersons' terms, Zhendong explains to me that his different companies can not only give advice and broker deals for the acquisition of aircraft, but can also directly operate aircraft for commercial GA purposes.

That a small consulting company founded in 1993 has developed into such a comprehensive aviation services and operations firm in just under two decades is nothing short of a miracle on par with China's economic expansion as a whole. Moreover, as is perhaps typical of the kind of entrepreneurship so widespread in China, Avion was built by someone whose credentials—an English degree earned in China and an MBA earned in the US—do not quite match what one expects in the aviation sector. Zhendong's success so far has come not so much from his formal academic training but rather from his ability to improvise, respond to changing circumstances, balance risk, and continually capitalize on opportunities that come up in China's reform process.

Of course the expansion of the GA sector depends heavily on the accumulation of wealth. China's reforms since 1979 have created three very different social strata, what Americans would term the lower, middle, and elite upper classes. At the bottom are the rural poor and millions of migrants into cities who now form an urbanized underclass. In the middle are the many urban citizens who have been able to participate in higher education and launch stable careers. Increasingly for them, the symbols of success include ownership of houses and cars. And last, far outstripping the others in terms of income, are the power elite, the extraordinarily wealthy few who have ridden the tides of reform to the top. According to *Hurun List*—a widely-quoted publication that tracks the wealth of China's super rich—China by the end of 2013 had 1.05 million RMB millionaires and 8,100 billionaires (the RMB, or "yuan," is the Chinese unit of currency). By 2015, *Hurun's* headline declared: "596 Dollar Billionaires in China, up 242, Overtaking US for First Time."[9] Half these people planned to purchase yachts, and one in six planned to purchase private planes, for doing so would represent a mark of success. The average age of these richest citizens was 43, roughly 15 years younger than the average "super rich" in the rest of the world.[10]

Caught up in the Four Modernizations

Zhendong has patiently waited and prepared for the arrival of this day. In many ways, his story started in 1979, when Deng Xiaoping ushered in the Reform and Opening Era with his call for attainment of the "Four Modernizations." At the time, Zhendong was a college freshman. He, of course, had no idea where the future would take him or his country. Upon graduation four years later, Zhendong, like everyone else in his graduating class, was assigned to a job by state officials. Zhendong went to the government's customs bureau in Guangzhou, but he soon quit that job and left for Shenzhen.[11] Still a small fishing town at the time, Shenzhen—along with Zhuhai, Shantou, and Xiamen— had just been designated one of China's first four special economic zones, experimental areas for market-based reforms.[12] In those early days, when Shenzhen felt like one big, dirty construction site, Zhendong became an aviation interpreter for pilots in helicopters. Yet his career did not take off just then.

Reminiscing about that time, Zhendong recalls some close

calls. In a voice touched with seasoned wisdom he says, "The S76A model A was what I flew in for 1,000 hours as the third crew member. That Allison engine was finicky like a lady." The unfailingly polite Zhendong pauses and adds, "Pardon my analogy." Then he explains that these engines caused quite a few fatal accidents in the mid-to-late 1980s. The aircraft operators had to change engines constantly, sometimes every month or two. While hanging out in bars with pilots in those days, he would often hear the names of individuals he knew who had gone down in crashes.

And the danger would creep even closer to home. Zhendong talked about the time he was awarded the CAAC's third degree honor for rescue. A workboat in the South China Sea had been listing 15 degrees in a strong typhoon, and, as Zhendong recalled, Esso's offshore exploration rig *Jim Cunningham* relayed the call for a helicopter rescue. When one of the helicopter's crew—a Communist Party member, no less—refused to go out in the storm, Zhendong, in addition to his onboard interpreting duties, took up the role of navigator and completed the rescue.

"I was 25 or 26 years old then," he explains, "an age when death was just a concept. Now my partners try to limit my rides on training and exciting fun flights," he chuckles. "They don't even want me to smoke. They want me to buy life insurance and have a physical checkup. They have invested in me, not just in my company."

It's difficult to imagine anyone would be concerned about his health by just looking at him. Just past fifty, a headful of silver hair frames his tanned face. He is practically free of wrinkles, and his eyes radiate a quiet self-assurance. Mostly because of his silver hair, but also his compact and trim physique, he has often been likened to the former Japanese Prime Minister Junichiro Koizumi. He is also an avid and accomplished golfer who proudly displays trophies and photos with golf stars in his home and office. But remembering his active and almost mischievous reputation in college, I could easily imagine a much younger version of him in the middle of the events he recollects. He says his earlier experiences made him acutely aware of the vast technical gap separating China from Western countries.[13] To help close that gap, he went abroad to earn an MBA so as to improve his ability to manage international collaborations.

While Zhendong was in the US, China's reform process, having completed its first decade, began to hit a series of crises. Indeed, the government's crackdown on the student movement in Beijing and other major cities in 1989 threatened the entire reform effort. Soon thereafter, the dissolution of the Soviet Union further tested the resilience of China's political system. As the 1990s progressed, concerns steadily grew over Hong Kong's impending reversion to Chinese sovereignty. In the years immediately leading up to the slated 1997 hand-over, some 60,000 Hong Kong citizens would emigrate annually.[14]

Casting His Lot with Modernization

To arrest sliding confidence in continued Communist Party rule and signal his own determination to stay the course with economic reform, Deng Xiaoping made two important moves. First, he saw to it that Hong Kong's pending reversion to Chinese sovereignty remained on track. Separated from Shenzhen by only a river, Hong Kong remained a British colony in the post-World War II era long after other Western nations had given up their colonies throughout Asia. If handled correctly, Hong Kong's hand-over would reinforce Beijing's claim to be the legitimate government for all of China while still affording the People's Republic a window onto the outside world through a separately governed territory. So even as the 1989 student movement rocked the Chinese capital, work continued between the United Kingdom and China on the drafting of Hong Kong's "Basic Law," the constitutional arrangement that would govern the territory's transfer of sovereignty.

The second thing Deng did was to undertake his now famous Southern Tour in 1992. The trip, highly publicized in the Chinese media, was intended to reinforce the status of the four special economic zones and maintain forward motion in the overall national reform effort. This was particularly important at that historical moment, for, in the aftermath of the 1989 student movement, conservatives in the Chinese government were substantially curtailing reform and threatening much worse. Deng's Southern Tour amounted to a kind of end run around those conservatives to break though obstacles they had created and get the change process going again. It was only a year after this important event that Zhendong returned to China from America with his MBA. He chose Hong Kong as the place from which to launch Avion Pacific. Hong Kong was not yet officially part of China, and precisely because of that Zhendong could take advantage of the well-established banking system that had grown up under British rule.

Casting one's lot with China's reform at this particular moment took a leap of faith. But Zhendong trusted that Deng would be able to rejuvenate the stalled change process. As Zhendong sees it, he entered China's GA market at an extremely bleak but oddly propitious time. The problems were myriad with everything from depressed markets to blocked skyways. However, Avion was entering a market with few if any competitors. There was no set map for a private company like this to follow, but that meant the possibilities were boundless. As Zhendong later remembers, "It took a creative mind and a lot of daring to find a way forward." And, as his business partner Bei states, "Zhendong found a niche."

Prior to the 1990s, China's aviation safety system was based on a model directly inherited from the Soviet Union. In the context of this existing system in the late 1980s, China's national airline, CAAC, was split into a series of regional players, thus creating 26 separate airlines by 1994. The proliferation of airlines and an increase in air traffic overtaxed the aviation safety system and a spate of accidents ensued. Some 400 airline accident-related deaths in the early 1990s served as a wake-up call for reform of the aviation safety and air traffic control systems.[15] While considerable attention was still devoted to military flights, new policies were implemented to improve coordination and safety in the civil aviation sector.[16] At least some people in the mid-1990s could see that China's aviation system would have to shift structurally. Elizabeth Keck, the FAA's senior representative in Beijing at the time, described change as "in the offing."[17]

Though 95% of the airspace was still under military control in 1994, the balance of power between CAAC and the People's Liberation Army Air Force (PLAAF) started to shift when CAAC was elevated to ministerial status. At the same time, CAAC began making headway in negotiations with the PLAAF to expand the amount of civilian-controlled airspace and upgrade air traffic control systems.

There is No Turning Back

Zhendong's entry to the aviation field was a process of on-the-job learning. Decisions were often made on an ad hoc basis with little time for deliberation. In the early days, the business mostly involved helping foreign firms partnered with Chinese counterparts in offshore oil exploration. The offshore rigs required helicopter support, so Zhendong helped foreign firms secure aviation fuel at the kind of domestic rates that only CAAC could normally get, access airport facilities that only CAAC could access, and assemble aircraft crews with the kinds of people only CAAC could normally hire. In Zhendong's words, "You [the foreign firm] spend the money, and I get you CAAC."

In 1995, Avion began selling used aircraft and spare parts. Two years later, Avion became the exclusive agent for Sikorsky helicopters. Its first customer was the Chinese government's transportation department. Later, Avion came to represent more companies and product lines, including McDonnell Douglas helicopters, Dassault Falcon Jets, Hawker Beechcraft, and Eclipse. As the business expanded, so too did the services offered. "Once the airplanes are in," Zhendong explains, "we provide our clients with service and management."[18]

As the business expanded and various market opportunities grew, Avion continually improvised to make things work. For example, all the way into the early 2000s, the military still made all airspace decisions along China's 1,122 civil airways. They made few if any concessions to GA traffic,[19] and what possibilities existed for accessing the airways were further complicated by cumbersome flight planning regulations. China's airspace, like that of most countries, was divided into different blocks controlled by different regional control facilities. So crossing different airspace blocks meant submitting multiple flight plans to numerous offices, an extremely time-consuming and complex procedure. Just delivering a helicopter to a commercial customer—even one who planned to use the aircraft for a highly localized application such as oil platform servicing—became onerous. Even getting the regulatory clearance to fly the aircraft to the customer was not without difficulties.

In 1999, three years into her tenure at Avion, Bei had to devise one impromptu solution while working as a coordinator for a Swiss geological team and its Chinese partner. They were developing gold mines in China's far western Xinjiang province and sought Avion's help in acquiring a helicopter from Hong Kong. But how would the helicopter travel over the many separate airspace blocks between Hong Kong and Xinjiang? At a loss for better options, Bei had the helicopter disassembled and shipped the parts by train. The regulatory situation was such that it was easier to ship a helicopter and 80 barrels of aviation fuel 3,000 kilometers rather than simply fly it.

Even then, the problem was not fully solved. At the time there were no rules for sharing airport facilities with outsiders, especially civilians. So when her technicians showed up in Xinjiang to reassemble the helicopter, Bei had to "rent" space in a local airport's hangar by handing over some "gifts." She then hired a caravan of trucks to transport fuel barrels to the gold prospectors' campsite, a one-day trip away in the mountains. As soon as the caravan got out of sight of the city, however, the seven truck drivers pulled over and collectively demanded a steep hike in payment. There was nothing for Bei to do but agree.[20]

Bei is tall, athletically built, and still the champion athlete I knew her to be in college. She has sported a short hairstyle for as long as I have known her. As quick to smile as to act, she has clearly taken that style into her work. Despite the ordeals she endured in the early days in Xinjiang, Bei saw even then that there was "clearly no turning back." Technology and GA were in China to stay.

Zhendong never thought much of the business practice

of describing long-term visions. In his view, opportunities arise not because of a particular company's vision, but rather through changes coming from society at large. In Zhendong's view, the key is to spot emerging societal needs and respond. With the dawn of the new millennium, Zhendong understood emerging opportunities that Avion could capture in relation to the need for special helicopter training for domestic clients, the development of standard international-level operating procedures for offshore drilling-related helicopter services, and the introduction of helicopters for power line maintenance in the electric power sector, among many others.

But Zhendong clearly has done more than just respond to market opportunities ad hoc. Though he lacked formal training as an engineer, Zhendong leveraged his other strengths, particularly his knowledge about and connections to the United States. As Zhendong saw it, the first step toward dismantling the military's monopoly over Chinese airspace was to draft Chinese civil aviation regulations on the American model. This impulse is evident in some of the first general aviation regulations of the time, including the 2003 "Regulations for GA Flight Control." However, this set of rules was still highly restrictive. It required a 15-day lead-time for the filing of a GA flight plan and it did not permit unlimited access to the low-altitude airspace (below 1000 meters) in which many GA aircraft operate. Once these rules went into effect, users still encountered uneven application across regions. Thus the kind of obstacles Bei faced in Xinjiang when seeking to ferry a helicopter across multiple airspace jurisdictions remained in place. But, the fact that these new civilian regulations had been issued with the support of civilian aviation authorities within the State represented an advance. The military had been forced to cede some authority and some airspace to civilians. And though the government was moving gingerly, it seemed to be signaling that more change would come.

Using his US connections proactively, Zhendong pushed for certain kinds of change in the Chinese market. First, he reasoned that for GA to develop in China, it had to acquire an independent status as a sector with its own rules and regulations. When he put CAAC in contact with Netjets, he was putting together a major national aviation management organization with the world's biggest aviation fractional ownership company. In fractional ownership, several different parties, whether individuals or companies, share ownership over expensive assets such as aircraft. Each owner owns a piece of the asset, and the owners collecively share use of the assets and shoulder their costs. The point here is that CAAC and Netjets together had an interest in seeing the GA market in China grow. The two organizations worked jointly to develop the Chinese Civilian Aviation Regulations (CCAR), including CCAR-91, "General Operating and Flight Rules." The CCAR was modeled on the FAR, the US's Federal Aviation Regulations. CCAR-91 paved the way for CCAR-135, "Rules on Operation Certification of Small Aircraft Commercial Transport Operations." Other rules that followed gradually made it easier for small aircraft to operate in China. With its early introduction to the Chinese GA industry, Netjets then gained a foothold in the market. It started to bring its business to China in 2011, setting up its own Aircraft Certification Office (ACO), and now works with Avion as a partner in selling Netjet Marquis Cards in China and Hong Kong. This fractional ownership program was scheduled to start around 2015. The establishment and the subsequent gradual relaxation of the rules and regulations for GA also expanded Avion's horizons for future development.[21]

Even today, Zhendong would like to see greater easing of the military's dominance over the air traffic control system. "I make noise whenever I can," he says, "whether at airshows or conferences or when meeting with leaders." But, he adds, he is also realistic in maintaining cordial relations with those in the existing power structure, be they military or political leaders. Zhendong notes with pride that given his experience and connections, approval for Avion's flight plans is only a phone call away, not the arduous seven-day process mandated by current regulations.[22] Veterans like Zhendong clearly know how to navigate the system.

"Everyone is Entitled to Live Life to the Fullest"

Days into my visit to Shenzhen, Zhendong gathers a few of our college friends for a chat in his office.[23] Zhendong and Sen, two men in polo shirts, sit in a pair of red-leather chairs and chain-smoke. Bei in her neat business outfit drops in from time to time while taking breaks from her regular workday responsibilities. She takes the chair by the door. I complete the circle by sitting in the cozy space next to Zhendong's oversized desk. Thirty years out of college, we could all still remember Zhendong's college pranks. We laugh and recall that cold winter day when he poured a bucket of icy cold water onto a guy singing inside an unheated shower stall, eliciting a howl that made everyone in the five-story dorm think a homicide had been committed. Yet in his CEO's office now, he has an air of authority and expertise far removed from the mischievous young man we knew in college.

Zhendong is not shy about reflecting on his leadership style and beliefs about civic responsibilities.

"The leaders sit right there and listen to me," he says, pointing to my seat next to his desk. Sen, by way of explaining, yet with a trace of envy, seconds his statement and adds, "They don't know his business." Sen implies that the leaders accept Zhendong as a greater expert about GA than themselves. "If you're running a business they do know, like my wholesale food company, they bother you all the time."

In a business in which no more than five percent of startup companies survive their initial years,[24] Zhendong attributes Avion's twenty-year longevity to three things: his leadership style guided by the scholarly Confucian doctrine of the Mean, his passion for aviation, and his craving for success and a good reputation. Pointing to the golf trophies standing among his photos with golfing stars and various leading lights in the aviation world, he confesses, "I don't like to lose when I play golf." As for his love of aviation, "what boy doesn't like airplanes?" he asked rhetorically. That love might also have come from his father's genes, he adds, but he couldn't know for sure because his father, a railroad engineer, passed away when he was five.

But it was the Chinese tradition based on Confucianism

and Daoism that spurred him on. "The Dao that can be said is not true Dao," he says, quoting the first line of the *Daodejing*. This turn of our conversation took me by surprise. In a span of minutes, we moved from talking about a college prank to business, and now, to philosophy. A business like his is built on trust rather than rules. Most of the deals he signs off on are in the millions of US dollars. "Trust is the reason they hand you the money," he says. Citing an example of the recent sale of a Hawker Beech business jet to a well-known billionaire named Chen Fashu, he concludes, "The whole negotiation took two hours. Few in this business could seal such a deal in that amount of time. He didn't even go through the contract. I told him, 'when you have to go search for a clause in the contract, it's already too late.'"

He gives us other examples. When buyers of Hawker Beech jets panicked at the company's Chapter 11 bankruptcy filing in 2012, Zhendong calmed them by explaining that King Air jets, a line of Hawker Beechcraft products, was still intact. He assuaged his clients' anxieties by promising to assume all responsibilities should their orders go bad. To be stuck with half-paid airplanes could cause liquidity problems that might be serious enough to bring down a company. Yet according to Zhendong, his confidence comes from a number of difference sources: his knowledge of the supplier, his understanding of the buyers' finances, the backing of his strategic partner, Seacor, and his solid performance record.

As if to illustrate his relationship with his clients, the phone on Zhendong's desk rings. A client is calling to thank him for his help getting over a financial crunch. "So here's a perfect case in point," Zhendong says as he hangs up the phone. "This client couldn't come up with the last payment," he explains. "This is what I mean by not playing by the rules and building trust at the same time. So what do you do in this case? Do you take legal action? What good would that do? Do you try to throw him in prison? No one wins in that case. So I worked out a plan with him. I'll take over his contract and pay the remaining bill, and he will pay me back as soon as he turns his business around." Of course Zhendong can't do this for every client, but in this case

Zhendong felt that he knew the client well enough to give it a try. As he says, "Now things have come through and I have won a loyal customer. Everyone wins."

The suspense and uncertainty in Zhendong's stories all boil down in his view to manageable risks. Yet from my perspective, I sensed that any of the major risks he was assuming on behalf of his clients could bring down his entire company. I ask why a client would stretch his finances so much to buy an airplane and why, in turn, Zhendong would stretch so much to back his clients. Extending oneself so far to pay for something that doesn't seem to have much practical use does not appear to be a sound business move.

"People are different," Zhendong explains as he lights another cigarette, "but we all work to attain maximum control and freedom in life." Clearly, I had just given him an opening to make a case for what he does. The advantages of or passion for owning a plane have to outweigh the financial burden in order for a client to make that decision, he says, warming to his topic. Then he effortlessly lists the advantages of plane ownership. With your own airplane, he begins, you save time and have more control over your own safety, as long as you hire a good pilot and have a good relationship with him. A plane provides privacy, which is hard to come by in China. And an airplane is an effective promotional tool. Owning a plane impresses business partners as evidence of solid finance, Zhedong argues, somewhat ironically, given his stories of assuming the liabilities of overstretched clients.[25] The high operating costs of aircraft ownership also provide opportunities for tax write-offs. These reasons roll off his tongue as though he has expressed them many, many times.

"What GA offers to a modern society is needless for me to repeat," he concludes. "Agriculture, search and rescue, geological surveying, EMS, pilot training, linking remote areas to the developed parts of the country . . . it is impossible to list all the amazing things small aircraft can do." At times a wise scholar who talks about Daoism and Confucianism and at times a confident businessman, he is now a consummate advocate for a modern civil society, where the freedom to choose one's lifestyle and apply advanced human technology to facilitate that lifestyle are basic prerogatives.

Clearly mindful of tensions between environmental concerns and the push for the kind of consumption he advocates, Zhendong is well prepared to respond: "I for one would also like to go back to a simpler and cleaner time. Yet, if the environmentalist's child has a medical emergency in the forest and only a helicopter can save the child's life, would the environmentalist deny that chance for the sake of the forest's wellbeing a thousand years down the road? My point is that we have this technology to make life easier, so why not use it? As for wealth, it endures through time. Wealth gives people time and leisure to create beauty in life, and beauty endures."

Bei walks in from the front office and picks up where Zhendong just finished. "I just returned from India," she says, "and, you know, I travel widely. Be it ruins or still-standing grand palaces such as the Taj Mahal or Versailles, you have to marvel at the past grandeur that still shines through. But, where are the huts of the workers who built those palaces? They are long gone." Zhendong finishes the thought for her, saying, "Everyone has his allotted time, and everyone is entitled to live life to the fullest."

Bei and Zhendong are not only business partners. Their two families often travel together and are very close. Zhendong's wife, Sheena, credits Bei with being the "Mrs. Capable" who takes care of the everyday operations, and Sheena herself, who used to work for international firms, helps with Avion's accounts from home.

While on one of their business trips, disaster struck at home. "Bei and I were half an hour away from Las Vegas when I got this call," Zhendong explains. "I was told, 'We've lost contact with a pilot for fifteen minutes.' It's a big deal for a helicopter to disappear for fifteen *seconds* from the radar screen, so you know what fifteen *minutes* means. I remember the dead silence as I hung up the phone." Bei cuts in, "We turned around right away, and headed right back to the debris from that storm over the Huangpu River in Shanghai. Two died." "But what I want to say is this," Zhendong resumes, leaning forward, "I headed straight to the hospital where one pilot's life was hanging by a thread. As

I comforted his wife, I promised her, 'As long as I'm alive, I'll see to it that your husband is well taken care of.' He's now living in Canada."Unfortunately, the other pilot didn't make it. Bei visited his widow in South Africa on her last trip, and the widow told her that she and Zhendong are her best friends. "It's important to remember that first of all we are dealing with people,"says Bei. Zhendong is quick to point out that the accident happened in 2005 and was the only crash of the 90 helicopters his company had sold by then.

This story brings Zhendong back to his reflections on the benevolent leadership style inspired by his understanding of traditional Daoist and Confucian philosophies. He takes another deep draw on his cigarette and turns to his relationship with his employees, returning to the topic of trust. "At every annual meeting, they talk to me about business, and I talk to them about Confucianism and Daoism," he says. "The best kind of leader is the person who does not let his presence be felt, and makes everyone take ownership of what he or she does. The second best kind is the one who is praised by people working under him. And the worst is the one who looks over people's shoulders." He says he gives his employees flexibility when family issues occur and fires abrasive managers. He then leans back in his chair and concludes, "Everyone has a role to play in society." For him, his role as the chief of Avion is to create jobs and to contribute to the GDP.

"Some Limits Can Be Stretched"
The next day, Zhendong takes our group of classmates to dinner. Afterward, we return to his home. It is an attached condo in the gated community on Turtle Mountain, a high-end residential area within 15 minutes' walking distance from his office. Entering the two-story-high living room, we see sets of rare wood furniture in classic Chinese style. Sprays of orchids and fine drawings by his three children grace the shelves and surfaces of the walls. Zhendong's only contribution to the décor seems to be the few golf trophies tucked in the corner of a shelf.

"I'm a bad dad," Zhendong confesses over tea, wine, and the

sweets his wife brings out. Pointing to a drawing by his second son, he says by way of illustration, "I violated traffic rules one time just to show him that rules are not the same as one's best judgment. And another time I kept driving when the gas tank's indicator flashed on empty, just to show him some limits can be stretched." One of his character flaws, he says, is that he always has to try new things.

This assertion is laughingly contradicted by Ouyang, their household helper. Zhendong is actually a creature of habit, she says. "He could eat noodles every single day!"

Someone else chips in with another example, "Or smoking non-stop."

"When did you pick up that habit?" another asks.

"The work stress," Zhendong admits.

His wife Sheena jumps in to say, "There is no night or day for him," reminding us of the 12-hour time difference between China and the US. "When he needs to talk to suppliers in the States, he has to make those calls at night from home. And the next day he's right back in his office."

Zhendong's oldest son is attending an American university, and his second son is preparing to follow suit. The second son is a sweet and talented teenager. He does not seem to be somebody who has received the extra dose of rule-breaking and rule-stretching gumption taught by his father. As the son shows me a collection of his intricate drawings of Harry Potter, dragons, castles, and mythical animals, I start to wonder about the legacy his father has tried to pass on to this young man. On the surface, all the talk of rule-stretching, rule-breaking, and rule-making seems but a mischievous diversion from the serious application of Daoist and Confucian doctrine—something for casual father-son bonding. Yet it bears a resemblance, perhaps, to Deng Xiaoping's pragmatic sayings about the era: "Crossing the river by feeling the stones," and "Black cat or white cat, the one that catches the mouse is the good cat." But Deng, of course, did not specify whether or not it is an acceptable practice to catch the mouse by bending or skirting the rules.

A leader of a company like Zhendong's may have more rules

to follow than many others. To use the sky, he has to deal with the military, and to operate and expand that operation, he needs to deal with civilian authorities. It is indeed daunting to navigate between these two behemoths, especially since clear boundaries between their respective areas of authority remain unclear. But he leaves it up to the two giants to work out their own deals while focusing instead on his company's growth. As he sees it, GA is right now on the verge of a new round of expansion.

The Last Frontier in the Reform

Many in the industry seem to see the same picture. "GA is the last frontier in China's economic reform," Bei says. According to the *People's Daily* overseas edition of November 30, 2012, the number of Chinese GA aircraft is still only just above a thousand, compared to more than 240,000 in the US and 10,000-plus in Brazil. Yet when a nation's per capita income reaches US $4,000, the article asserts, GA will move onto the fast track for development. The article quotes the IMF's 2011 figure of $5,414 for Chinese per capita GDP. China's latest five-year plan (the 12th, for 2011–15) puts GA on the reform agenda and the CEO of the China International General Aviation Corporation, Qu Tiecheng, estimates that GA will increase at an annual rate of 20%.[26]

Honeywell's Asia Pacific CEO, Rishiraj Singh, was equally optimistic when he claimed in 2013 that for the next 20 years the spotlight of GA's growth will be on China's helicopters and business jets, the number of which will most likely increase tenfold.[27] A big part of Avion's business is helicopters. "Right now," Zhendong says, "the military has over 900, and civilians well over 200. Three years ago in 2010, we held a dominant 53% share of the helicopter market. This year we expect to be first again."[28] Avion also represents business jet manufacturers. This is one of Avion's later additions, since the government has only allowed individuals to own and fly private aircraft since 2003. And business jets have also been on a steep upward trajectory in the overall market. "In less than a decade, the number of mainland-registered business jets has rocketed from zero to more than 130," according to an online article on *CNN Travel*.[29]

The ultimate regulator of China's GA industry, however, will be the Chinese government. This has become clear during President Xi Jinping's anti-corruption campaign in 2014, which caused a drastic dip in the predicted steep rise in the GA market. Bradley Perrett, a columnist for *Aviation Week*'s online edition, points out, "A big question is when the strong rise will begin, and it depends mostly on Chinese politics."[30] Given the military's and government's tight control, and the many strict regulations for the operation of private aircraft, many private owners take to the skies in secret and illegally, giving rise to the term "black flights." At the current price of between RMB 10,000 and 50,000 (about US $1,700–$9,000), depending on the region, the fine for an illegal flight—when the violator is caught—has seemed to be an affordable price to pay for many owners.[31]

If taking to the skies is a basic human dream, that dream has yet to find a comfortable landing pad in communist China. Though the environment is still rather forbidding, many now seem willing to pay to taste that dream. Thirty years of sustained economic growth has clearly provided some citizens with the means to do so.

For Zhendong, "Aviation is my dream." He continues, "I've gotten offers from American companies, and there was at least one that seemed very, very attractive. But, you know, all this is not so I can make enough money for an early retirement. I'm building and pursuing my own dream."

Indeed, he has a big dream with many pieces in place already, and many, many more to come. Establishing King's Aviation was just one important step toward getting a share of the expanding market. King's Aviation has also partnered with AVPRO, the world's largest broker of second-hand airplanes, to cater to a global clientele. And, finally, to facilitate the flow of aircraft across national borders, Zhendong has created yet another company, SHR, to deal with customs authorities and arrange import and export procedures.

"Zhendong has an artistic bent, and he has a vision," Bei says of her business partner. And Zhendong admits he does keep a long to-do list with items now at various stages between reality and aspiration. For example, Avion has purchased a piece of land at Zhuhai Airport to start a maintenance center. He wants to set up an MRO—a maintenance, repair, and overhaul operation (also know as a Part 145 operation)—to provide services for GA operators. He will also continue to help strengthen China's aviation codes. He indicates that "Boeing helped China write the CCAR-121 in order to sell their planes to China and help China adapt to international flight rules." He adds, "We helped in CCAR-91 and -135. The next thing we want to do is to help set up HIA—a helicopter industry association—in China." Zhendong takes his business models from America and says the accompanying aviation codes are a necessity for safe operations and rational sectoral expansion. His company used to invite foreigners at high prices to translate and compile the Chinese versions of safety regulations. Safety for him is an issue of social conscience and moral obligation.

Setting up an aviation school is another item on the list.[32] Avion has already represented two large US advanced training companies to arrange crew training in America for Chinese law enforcement and search and rescue (SAR) agencies. He says he wants to bring that special training to Chinese soil. Further down the road he wants to facilitate GA financial leasing and perhaps create a GA fund that invests in various sectors. And his list goes on.

"With China at my back, the foreigners have to deal with me," he says. "If I'm in America and work for them, they don't have to take me seriously. The country is my strongest backing." And there is every indication that the country will continue to back him up. When Xi Jinping assumed his position as China's leader in 2013, the first trip he took was to retrace Deng Xiaoping's Southern Tour, a clear statement, in Zhendong's view, that Xi wants to continue what Deng started.

Outside the window of Zhendong's office, the 30-year-old city of Shenzhen looks like a city that has always been there. Mature trees line the boulevards, and the neat buildings appear weathered by the subtropical sun and rain. Over a century before, Shanghai had risen from a similar status as a small fishing village and in a short period of time it has become a major cultural and commercial center in the Far East. The grand mansions lining Shanghai's Bund along the Huangpu River today attest to the grandeur of its colonial era. But unlike previously foreign-dominated Shanghai, Shenzhen is an entirely Chinese creation and Zhendong was there from Day One. This once small fishing village has become the launching pad for China's economic miracle, and Zhendong is riding its meteoric ascent.

Chapter 3 Notes

1. In an article "Spreading Its Wings," *China Business Review*, Vol. 22 (1995), Elizabeth Keck, senior FAA representative in Beijing at the time, writes: "Unlike the US, where all airspace is managed by the Civil-run FAA, more than 95% of China's airspace is controlled by the military and reserved for military purposes." She further explains the military management of the airspace in another article in the same magazine in 2001, titled "China's Changing Skies": "Until recently, China's military made all airspace decisions about traffic in its 1,122 civil airways. The military, which had ultimate authority over any changes in routings, flight clearances, and other routine activities, relayed these decisions to China's civil air traffic authority—the Air Traffic Management Bureau (ATMB) under the CAAC. ATMB controllers, in turn, passed the messages to the civil aircraft." *The China Business Review*, Vol. 28 (2001). Because direct Chinese sources of the early Reform Era are not readily available through public channels (even though the facts are widely known), I have resorted to using the publications of official US representatives.

2. This is quoted from a news article by the customarily unnamed "spokesperson" for this online publication of the Ministry of National Defense of the PRC titled "Maintaining Airspace Security is a Common Responsibility of the Entire Society" (see http://news.mod.gov.cn/pla/2014-02/25/content_4492494.htm). The General Flight Rules of the People's Republic of China 中华人民共和国飞行基本, first compiled in 2000 and modified in 2007, also make clear that the priority is national security. The first item under the second chapter, "Airspace Management," says: "Airspace Management should maintain national security . . ." The third chapter, "Flight Management," starts with a similar emphasis: "The flight management within the border of China is uniformly executed by Chinese People's Army Air Force . . ."

3. In the "Temporary Regulations on the Management of the General Aviation 国务院于通用航空管理的时定" issued by the State Department on January 8, 1986, the new term "general aviation 通用航空" replaced the old "specialized aviation 业航空." General aviation shared its rules and regulations with civil aviation until the State Department issued "Regulations on the

Flight and Management of General Aviation 通用航空飞行管制条例"on January 10, 2003 (see http://www.caac.gov.cn/H1/H3/). From the inception of the Reform and Opening at the end of 1978, it took general aviation eight years to gain its own status with its current name and own set of operational regulations. It took ten more years to implement a set of more comprehensive operational rules: "Regulations on Approval and Management of General Aviation Flight Plans 通用航空飞行任务批与管理"on December 1, 2013.

4. The United States has always been the ultimate point of comparison in discourse on the development of China's general aviation. The US is the goal China is heading for. US statistics are everywhere in the publications about general aviation. This mentality to a large degree resembles that of "Surpassing Britain and Catching up with the US," as in the '50s and '60s when the Chinese campaigned to push up their GDP. At a site for professional civil aviation articles, for example, one finds a typical article in "60 Years of CAAC: A Brief Introduction to the Development of China's General Aviation 民航60 年：中国通用航空发展概况."The section called "The Gap Between China and Nations with Advanced Aviation" contains a few charts. The nations in the chart of comparison include: the US, Canada, Australia, and Brazil. Specifically, the article contains a more detailed chart comparing China with America with these items: total hours of flight, total number of users (enterprises and individuals), the number of certified pilots, the size of the fleet, and the number of airports. Although this article was written and posted in 2009, all of these same numbers and categories have been cycled and recycled over the years since. They have been updated, commented upon, and compared with China's rising number to show a slowly narrowing gap between the two countries (see http://news.carnoc.com/list/145/145212.html).

5. I spent a week with Zhendong and his family in Shenzhen during the Spring Festival vacation. Zhendong's company, Avion Pacific, kept a regular work schedule during that week, allowing me to participate and observe many of their routine operational activities. My interviews and other related activities during that stay provided the main basis for this story about Zhendong and Avion Pacific.

6. The S92 helicopter apparently comes with many types of fittings so that the same airplane may serve different functions. On the Sikorsky website, it offers different "missions": Offshore Oil, HOS (Head of State), Executive, EMS, SAR, and Airline.

7. As Elizabeth Keck, FAA senior representative in Beijing, writes: "China has developed an impressive indigenous training capability for both pilots and flight attendants" ("Commercial Aviation Takes Off," *China Business Review*, Vol. 28 [2001]).

8. In 2010, the PRC State Department and Central Military Command jointly issued "Opinions on Deepening National Low-altitude Airspace Management and Reform," indicating the government's attention, as well as intention to fast-track the development of general aviation.

9. The exchange rate in 2013 was about $1=6 yuan, and about $1=6.2 in 2015.

10. See "2010 Hurun Wealth Report: China's Millionaires Number 875,000" 2010 胡富报告：中国千万富豪达 87.5 万人 (see http://finance.sina.com.cn/leadership/crz/20100401/15347678221.shtml).

This growth in wealth among China's richest continues, as reported in a 2015 report in *Aviation Week Network* that quoted Airbus: "By 2017, China should have almost as many billionaires as the US . . . relying on studies by Ledbury Research" (see "Bizjet Market Shows Faint Signs of Recovery in China" at http://aviationweek.com/business-aviation/bizjet-market-shows-faint-signs-recovery-china). In another 2015 *Hurun* article, the title says it all: "In the Next 3 Years, the Number of Chinese Millionaires will increase to Around 1,210,000." See https://www.google.com/?gws_rd=ssl#q=number+of+hurun+2014+millionaires. *Hurun*'s in-house survey culminates in the following report at http://www.hurun.net/en/ArticleShow.aspx?nid=262. In it, *Hurun* postulates that 40% of China's super-rich "plan to use private jets, half of whom plan to time share."

11. For the government's post-college job-assignment policy, please see note 8 in the Introduction.

12. For the Reform and Opening, see note 2 in the Introduction. For the Four Modernizations, please see note 13 in the Introduction.

13. In *China Airborne: The Test of China's Future* (Vintage, 2012), James Fallows recounts an aviation detail related to Kissinger's first visit to China in 1971, paving the way for Nixon's visit the following year (an anecdote he learned from a friend in the PLA Air Force: Kissinger took a Boeing 707 from PIA, Pakistan International Airlines). "At the time," Fallows wrote, "707 was one of the most recognizable aircraft in the world" (p. 45). But it did not normally fly into China, as its airports were closed to most Western airlines. So the technical challenges of the arrival of the Boeing 707 were perhaps the first China had to cope with: how to service the plane, and at a "more basic and potentially embarrassing level, how was Kissinger supposed to get from the airplane onto the ground?" (p. 46). The solution, as Fallows goes on to explain, was that instead of buying, or even borrowing, a standard airport staircase that would reveal their technological isolation, they built one and rolled it out at the arrival. Apparently, it worked.

A decade-and-a-half later, Fallows "first saw airplanes from China's old fleet in 1986 . . . Tickets were written out by hand . . . The planes were mainly old Soviet junkers . . . The plane's (Boeing 707) first flight, in 1980, was a point of national pride" (pp. 65–66).

Fast forward to 2009: a comparison chart in the article tracing the 60-year history of China's general aviation (see note 4 above) showed the following numbers:

Total flight hours: 260,000 hours in China, 27,705,000 hours in the US
Number of operators: 74 in China, 15,000 in the US
Number of licensed pilots:
China: 2,237 for GA, 1,409 private planes;
US: close 70,000 in total

GA fleet: total of 801 in China; 22,200 in the US

Airports:

China: 398 total, with 69 GA airports and 329 temporary landing sites

US: 18,000 airports

(http://news.carnoc.com/list/145/145212.html)

14. See Asia Pacific Migration Research Network (APMRN), a UN educational, scientific, and cultural organization: "Migration Issues in the Asia Pacific: Issues Paper from Hong Kong" (http://www.unesco.org/most/apmrnwp7.htm).

15. *The Southern Weekend* 南方周末 (May 16, 2002) has an excellent long article: "Why Did Airline Disasters Happen One After Another? Explaining the Nepotism That Tripped the CAAC 空何接踵而至？解倒中国民航的裙系." It sketches the three stages of the CAAC reform and explains that, despite the reforms, nepotism within the system prevented the reforms from effective implementation. It points out that the general aviation industry should be run as an enterprise, separating business from politics.

In her article "China's Changing Skies," Elizabeth Keck also points out that the "CAAC had discovered that rapid expansion had pressured airlines to promote pilots with insufficient flight experience to captain too quickly." So, Keck explains, Order 77 was drafted to require "pilots to fly a minimum number of hours (based on plane size) before they can be promoted to captain. Its requirements are even stricter than those set by the International Civil Aviation Organization and the US FAA." She attributes Order 77 to being the key reason for China successfully reducing the accident rate (*China Business Review*, Vol. 28, 2001).

The three stages of CAAC reform are divided as follows:

Stage One (1987–1994): This period initiates the decentralization process. It separates the business operation from the political administration, making managing bureaus, aviation companies, and airports separate entities. It separates general aviation from civil aviation, reorganizes managing entities into smaller groups, and establishes new local aviation companies.

Stage Two (1994–1998): Specific measure are taken to implement reforms in civil aviation, to give more autonomy to business management in order to further strengthen airport and management systems. For this purpose, a new management system is established to connect local operations with the CAAC. In 1994, the CAAC Air Transportation Bureau 中国民航空中管理局 was set up, and in 1996, six major monitoring centers were set up; also in 1996, the flight paths of Beijing-Shanghai-Guangzhou were handed over to CAAC management.

Stage Three (1998–): This stage was still just in its early phase at the time the article was written in 2002, and the article did not lay out a road map from that point on. The highlight at the time was the conference in 2001 that brought together the enterprises directly under the CAAC to discuss reform and reorganization. This set in motion the heart of the reform on all fronts.

This information can also be found through many other sources. But the particular article about the airline disaster can be found at the following link: http://news.sohu.com/34/64/news200886434.shtml.

16. Elizabeth Keck, "China's Changing Skies," *China Business Review*, Vol. 28 (2001).

17. In "Spreading Its Wings," *China Business Review* (July/August 1995).

18. "The growth of the commercial aviation activity has driven the growth in aviation services. In the world market, for every \$1.50 spent on new aircraft, approximately \$2.70 is spent in related services such as maintenance and training" (Elizabeth Keck, "Commercial Aviation Takes Off," *China Business Review*, Vol. 28 (March/April 2001).

19. This number is from both Zhendong and Elizabeth Keck, "China's Changing Skies," *China Business Review*, Vol. 28 (March/April 2001).

20. Bei told me this story when she travelled to Beijing for business in the spring of 2013. We met in her hotel, and took a long walk in the city as she recounted many of her adventures.

21. The focal point of the relaxation of rules is the use of the low-altitude airspace, defined as being below 1,000 meters. It is the space most frequently used by general aviation. In November of 2010, the State Department and Central Military Committee jointly issued "Opinions on Deepening the Reform of the Management of Our National Low-altitude Airspace 于深化我国低空空域管理改革的意见" (the full document can be found, among many places, at this site: http://www.eeo.com.cn/industry/shipping/2010/11/16/185929.shtml). At the time when I interviewed Zhendong in early 2013, the excitement caused by this document was still very much palpable. Its first section clearly states that this reform "is a natural demand arising from socio-economic developement." The main objective, as stated, is "Through 5–10 years of comprehensive deepening reforms," to build a standard legal operating and service system. It lays out three major stages of reform: In the first experimental stage (before 2011), the reform would be implemented in a very limited number of cities; the second stage (2011–2015) would enlarge the scope of this experiment to more cities while fine-tuning the implementation, and the third stage, in 2016–2020, will be full implementation. It is notable that the second stage coincides with, and is part of the aviation reform in the country's 12th Five-Year Plan (2011–15). The significance of being part of the national plan is that it will be one significant portion of where money is directed.

22. In *Forbes Asia*, the online magazine (4/02/2014), Wu Zhendong is quoted as saying: "Ten years ago, it took 15 days to get approval. Then it was 7 days. Now it's one day or 2 days."

23. During my weeklong visit in Shenzhen, Zhendong and Bei completely opened their offices to me, allowing me to participate in their routine as well as special activities. The interview with the prospective pilot earlier in this article, for example, was an exclusive event with only the top officers present. My two hosts also gathered friends, including our old mutual friends, to discuss various issues. This was one of those social occasions. Aside from this

weeklong visit, I have in many ways and on numerous occasions benefited from their help in understanding the intricacies and technicalities of the GA business: emails, phone calls, and even their business trips to Beijing where I stayed at that time. They have been most patient and generous in sharing with me their knowledge and stories.

24. This is the number Zhendong gave me.

25. Gao Yuanyang, a Beihang University professor, wrote in one of his blogs that a "Business jet is considered as a big luxury toy to showcase status and extravagance. The wealthy people's attitude of 'ask not the price but the brand' both excites and puzzles the makers of high-end business jets." He then went on to explain the American saying of "No Plane, No Gain," and that 90% of the Fortune 500 use business jets (http://blog.sina.com.cn/s/blog_593827e90100x4p7.html [2012-3-13]).

26. Source: *People's Daily* overseas edition (November 30, 2012): "Pushing Open Low-altitude Airspace Across the Nation Next Year, Expect to have a Huge Big 'Cake' Dropping from Sky 低空放明年起全国推广，天上有掉下万亿大'蛋糕'"(see http://finance.people.com.cn/n/2012/1130/c1004-19746234.html).

27. Singh is quoted as saying, "The Chinese helicopter industry is in its infancy. It is still growing, so I think we are going to see a huge growth in the aviation sector over the next two decades." There were about 200 civilian helicopters in China when the article was written, and Singh predicted: "In the next two decades, there could be about 2,000 helicopters in China. You can see the huge growth potential" (see "Chopper Market Set to Take Off" *China Daily* (Europe), http://europe.chinadaily.com.cn/business/2013-04/17/content_16414510.htm [April 17, 2013]).

28. These are the numbers that Zhendong quotes from memory during an early 2013 interview. A different source puts the total number of military helicopters at 914 in 2013: "The 2013 issue of *The Military Balance* gives a total of 914 ... despite the lack of an authoritative number, a significant increase from the 500 helicopters listed during rescue operations in the wake of the 2008 Wenchuan eathquke (*China Daily*, July 5, 2008)" (Peter Wood and Christina Garafola, "Counting Z's: The Gradual Expansion of China's Helicopter Force." *The Jamestown Foundation, China Brief*, Vol. 13, no. 8 (April 12, 2013) http://www.jamestown.org/single/?tx_ttnews[tt_news]=40724&no_cache=1#.VvSRQik_o7A (accessed March 2016).

For the number of civilian helicopters, *China Daily* (August 9, 2013) has an article "Sky's the Limit for Whirlybirds" by Chris Davis, who quotes the number 298 from the Chinese Aircraft Registry for the year 2012. Asian Sky Group website (with Avion Pacific as a major founder) states: "As of year-end 2013, the Greater China fleet numbered 465 Civil helicopters, with 424 based in Mainland China, 30 in Hong Kong and Macau, and 11 in Taiwan" (see http://www.asianskygroup.com/attachment/news/ASG-THE-GREATER-CHINA-CIVIL-HELICOPTER-FLEET-REPORT-YE2013-en.pdf.

29. See Tracy You, *CNN International* (April 11, 2012): "Flying Dragons: Private Jets Are New Status Symbols in China," http://travel.cnn.com/shang-

hai/life/chinas-business-jets-reach-new-heights-113799. Also see Marvin Cetron, "China's Economic Growth Opens Skies for Bizjets," *Professional Pilot* (March 25, 2015) http://www.propilotmag.com/archives/2011/Oct%20 11/A2_China_p1.html. Cetron quotes Honeywell's estimation that China would buy about 100 business jets per year for the next five or six years. And "Boeing estimates that China will spend $600 billion on 5000 new planes in the next 20 years, including bizjets and airliners." Even the Vice Minister Wang Changshun of CAAC says that "by 2025 the country would have 3000 civil aircraft." And "The Greater China Business Jet Fleet Report Year End 2013" by Asian Sky Group reported: "The Greater China business jet fleet grew rapidly over the last two years, rising from 203 business jets in 2011, to 371 aircraft as of year-end 2013. From 2007 to 2013 the Greater China business jet fleet has grown at a Compound Annual Growth Rate of 34%, which is significantly higher than the global rate of 5%. Of the 371 aircraft, 248 are based in Mainland China, 97 in Hong Kong, 15 in Taiwan and 11 in Macau." Last accessed in November 2015. https://www.nbaa.org/news/2014/ ASG_Fleet_Report_2013.pdf. In a recent online article in *Aviation Week* (2015), "Bizjet Market Shows Faint Signs of Recovery in China," Bradley Perrett writes that after the bizjet market hit what he called the "brick wall" of the anti-corruption campaign in 2014, leaders of major manufacturers expressed optimism for a recovery in 2015. "In 2012, the consultancy McKinsey & Co. estimated the number of business aircraft in China would grow by 30% from year to year, but in 2014 Chinese customers ordered maybe half as many aircraft as in 2013 . . . And whereas McKinsey estimated in 2012 that annual fleet (and therefore delivery) growth would step up to 40% after 2015, there is barely a hint of anything like that right now . . . A big question is when the strong rise will begin, and it depends mostly on Chinese politics . . . The market appears to have recovered just a little, says Jeffrey Lowe of Asia Sky Group . . ." (http://aviationweek.com/business-aviation/ bizjet-market-shows-faint-signs-recovery-china).

30. Bradley Perrett, "Bizjet Market Shows Faint Signs of Recovery in China" (http://aviationweek.com/business-aviation/ bizjet-market-shows-faint-signs-recovery-china).

31. This information was given to me by Wu Qiongyao, the head of Avion's Shanghai office, when I interviewed her by phone in 2013. By 2015, official sources had increased the fine to US $32,000 (General Aviation Net 通航业网 [March 12, 2015], http://www.ethcy.com/htmls/info/thbk/thb-kzs/15513.html).

32. Pilots were already in short supply when I interviewed Zhendong in early 2013, and many more will be needed, as *Asian Aviation* magazine says (December 10, 2012): "The Asia-Pacific region alone will need hundreds and thousands of new commercial airline pilots over the next 20 years in order to support the fleet modernization plans of carriers in the region and the rapid growth in air travel, according to Boeing. The manufacture's 2012 Pilot and Technician Outlook predicts the region will need 185,600 new pilots through to 2030, with China having the biggest requirement—71,300 pilots" (http://

www.asianaviation.com/articles/364/Pilot-Training).This is close to three thousand pilots being trained every year. Right now, the CAAC's annual capacity of training is 2,000 pilots, according to another article in *Forbes Chinese* (September 29, 2013), "General Aviation, Next Trillion Dollar Market?" (http://www.forbeschina.com/review/201309/0028629.shtml).

CHAPTER 4

"Peek-a-book, Peek a World": Scientist Moms Championing US-style Children's Learning in China

Halloween

During the week before Halloween in 2012, Luo Ming and I walk into a bright and colorful children's activity room in the Wanliu branch of the Peekabook library. Wang Yi comes up to greet us. After a round of hugs, she says in her soft and unhurried voice, "Aidong is here today." Hearing mention of her name, she raises her hand with an enthusiastic wave and fades back into the circle of library department heads who are sitting around the low children's table. Someone passes around a bag of chestnuts over the laptops and notebooks. The paper goblins and pumpkins that hang down from the ceiling stir from the wave of excitement, and come to rest as the women start their weekly meeting. My friends had invited me to join them after I expressed interest in their work.

Wang Yi, Luo Ming, and Ning Aidong are the co-founders of Peekabook House. All three live in Beijing. The fourth co-founder, Hu Birong, has moved back to the States and still makes frequent trips to Beijing. Peeka, as they call the library, is their baby. Since its founding in January 2006 as the first non-government, non-profit private library in the country, it has grown to include four branches around the city with a new site in preparation. The concept has sparked much interest from both the media and the public, and Peeka has inspired the founding of dozens of similar libraries around the city.[1]

"Those candies are to be distributed, right?" Wang Yi asked in a matter of fact way as she points to a row of large bags leaning against the wall. She was clearly just checking the arrangement.

Seeing my shock at the amount of candy, Ming explains in a low voice that since the idea of distributing candies is still new in China, they usually buy a lot of candy and deliver it to the homes of people who have agreed ahead of time to participate. Then, on Halloween night, chaperones take groups of children trick-or-treating as they go door-to-door to those homes.

As I look through the glass door of the meeting room, I see a few small children climbing up and down the colorful blocks on the reading room rug while others are checking out books on the similarly colorful bookshelves. A mother is trying to interest her toddler in *What Do People Do All Day?* by Richard Scarry. Any American who browses the low shelves will find his or her childhood favorites in the original American editions, from stiff baby books to pop-up books to the *Magic Tree House* series. This is a comprehensive showcase of Americans' favorite books for preschool through high school children. Seven years after it opened, the library has gained over 2,000 regular members and amassed 90,000 carefully selected Chinese and English titles.[2]

Peeka remains the longest-standing, and most solid and influential library of this kind in the nation's capital. The trend it has started has made The Capital Library take note, and its president announced in 2010 that "The Capital Library has been paying attention to private children's libraries and is working on how to support them."[3] The support for the non-profit libraries that he pledged included making book donations and sharing its expertise. Ironically, the expertise and the kind of donations that the traditional library can offer were the very reasons for the existence and staying power of Peeka. Peeka provides not only thoughtfully chosen books for the youngest group of readers, but also a friendly environment in which they can move and learn, neither of which are available in public libraries.[4]

With this set-up and the collection, Peeka offers a philosophy of learning different from the currently prevailing trends. The founders say that to learn is to explore and to discover, and that learning should be fun. Unlike the traditional local schools and after-school classes that push Chinese children up the examination ladder, and unlike traditional Chinese stories that more often than not assume a didactic tone and deliver a social sermon, Peeka offers Dr. Seuss and Richard Scarry. Addressing the lack of quality picture books for young and very young children, it offers American-made, thick cardboard books for the youngest first-time readers. To augment early Chinese children's literary masterpieces that were clearly byproducts of adult literature titans such as Lu Xun and Ba Jin, Peeka offers works by the greatest professional children's writers in the West. It creates a world where children are at the center, where their dreams, silliness, and exploration and discoveries of themselves are the focus. The subsidiary status of children's literature in China stops at Peeka's door.

This drastically different library style has prompted the media's fascination with the illustrious educations and professional standings of Peeka's founders. The media expresses near incredulity that these women have given up lucrative careers and returned to China to tend a field that serves a slice of the population traditionally regarded as more infantile than intelligent.[5]

The Sea Turtles

These Peeka founders are popularly known as "sea turtles," a jocular twist on the translation of another word with the same pronunciation, *haigui*, "returning from overseas." Media reports never fail to emphasize the academic and professional backgrounds of these founders: Three of these women graduated from the nation's top engineering school, Tsinghua University, and the fourth, Hu Birong, from the nation's other top university, Beijing University. For many parents in China, the only measure of success for children and teens is to get into the best universities, and the golden diplomas of these four women are the ultimate achievement. But these women have moved far beyond that. The three Tsinghua graduates were part of the 13 percent of women in their respective electrical engineering classes by their own calculation.[6] After graduating from Tsinghua, Aidong's advisors were so impressed by her potential that they arranged for her to continue her graduate studies with their friends, well-known professors at UC Santa Barbara. Ming went on to receive her master's degree in China, and Yi pursued her master's in the States.

In the mid-to-late 1990s, when studying and living in the US was still a distant dream for even elite students across China, Aidong continued to a PhD program at UCSB. While Ming was visiting her husband in California, she underwent three rounds of intensive interviews and clinched a job at Stanford's GPS lab. Yi started working at GE as an engineer after finishing her master's degree, and Birong earned her MSEE degree after receiving her PhD in biology at UCLA. All four women married Tsinghua graduates who were working on their Silicon Valley start-ups. By the time these four women started their families in the Bay Area, Aidong had also landed a job at Cisco. To many, they had captured the American Dream. A children's library such as Peeka was never part of their career plans as they moved upward in their respective trajectories.

But their settled and thriving lives and careers were interrupted when their husbands decided to sell their start-ups and to return to China to set up new ventures. This decision meant their husbands would leave their supportive wives and children in the US. Although they were capable and independent, these young mothers soon felt the strain on their families caused by long periods of separation each year: They were raising their young children in America on their own while holding down demanding jobs. While one option would be to return to China and join their husbands, this posed other problems for each of them.

Ming's internal storm broke on a summer afternoon at an outdoor music festival with her friend, Hu Birong, who was about to leave for Beijing with her husband. "There I was," Ming said to me, "with a dream job as an engineer in a Stanford GPS lab at my favorite jazz concert with my good friend under the open California sky while across the ocean thousands of miles away my husband was starting his own company with Yi's husband in a rapidly growing economy. At home with me were our two-year-old twins, a boy and a girl, who were contending fo.' their own gender territory every day as their strong personalities emerged." Ming recalled how her life was being pulled by the two worlds and all things dear to her.[7] Yi had returned to Beijing with her family by then. Her decision had been made easier because she could transfer internally within the GE system where she worked. But Ming's job was not easily transferable. Nevertheless, she realized that a separated family was not a sustainable situation. She decided she needed to give China a try.

But what would they do in China? Be housewives? Both Birong and Ming wanted to find something meaningful to do if they could not continue with the current jobs that they liked. They reasoned that if they were to start anew, they should do something completely different. China did not need just another scientist. Their husbands were already contributing their expertise. The women wanted to take this brand new opportunity and make a difference. As new mothers returning to China for the main reason of raising a family, their thoughts turned to children-related ventures. After much brainstorming and research, they decided to transplant their favorite hangout place in the States to Beijing: the children's section of an American public library.

In 2005, Ming took a year's unpaid leave from Stanford to test the waters in Beijing. There she met up with Yi, who had been in Beijing for a year by then and was pregnant with her third child. She liked the library idea so much she never returned to GE after the child was born.

With Yi and Ming already committed to the library idea, Birong decided to take the next step. As Ming said, "Birong is always quick to take action." In evidence of this observation, Ming recalled that, on one summer day while driving around looking for a library site, Birong found a sunny spot for rent in a shopping mall and rented it on sight.

With a site secured, they decided it would be an English-only library to provide an English-only environment, and that they would pool together all the books they had brought back from America before their initial shipment arrived. When Aidong returned in 2006, the nascent Peeka library was unstaffed and badly in need of someone just like her—a hard worker who always did a good job. At this perfect after-school place for her children in Beijing, Aidong reconnected with her three friends. But little did she know their friendship and the place they set up would transform her own Beijing existence.

Urban Oasis

When I visited Beijing in 2009, my friend Li Feng proudly invited me to visit what he promised would be a "unique" library in the city. A successful entrepreneur in Silicon Valley, Feng told me he had returned so he would be able to tell his children he was there at the start of China's economic takeoff.[8] To me, that was a unique reason to return to China. As we drove in Feng's car to a forest of high-rises in the city and walked up to a second-floor suite in one of the massive modern buildings, I wondered what kind of uniqueness I was to find in this library.

There, I met Feng's wife, Yi, for the first time. Yi had just finished hosting that day's story hour. A slim woman of modest height and in her late thirties, her perfect almond face was framed by soft, shoulder-length hair. The features on her porcelain complexion were so delicate a violent laugh might easily shatter them. Her disarming smile and the words that flowed like a meandering stream defined the atmosphere as she showed me the library's operations. A group of children around a far table in a bright alcove were creating their own images from the story she had just read to them. Most of the kids were able to walk here from their homes, she explained, because the site of the library had been chosen to target densely populated areas and benefit from heavy foot traffic. In the beginning, she said, the founders had handpicked each and every book in the collection, and most of the books were in English.

As I looked at the familiar English children's books on the colorful child-sized shelves, the uniqueness of the library was somewhat lost on me, as I was then raising my own young children in America. The library felt to me to be just a very friendly, well-tended, and upscale version of an American library's children's section. But I soon realized just how Americanized I myself had become, recalling how, as a child, my mother would read me stories by the renowned children's writer Big Xin. A professor of children's literature, my mother would be engrossed in her reading as I started to fold paper animals. Of course, Mother was disappointed in me when she realized what had happened. Even if you don't like the story, she admonished, you should separate playing from reading. Libraries of my own childhood—the few times I was allowed in—were all hulking shells of silence.

By the time I visited again in 2012, Peeka's collection was about evenly split between books in English and Chinese, the result of local demand. Activities in the library had also expanded and diversified. Storytime increased and arts and crafts activities had been added, both free to members. Fee-based programs had also been added to generate revenue: English classes for different skill levels; excursions into the city, ranging from seasonal educational walks to concerts; summer and winter camps; holiday celebrations; and lectures for parents on family and parenting given by invited guests.[9] Many of these programs were free because invited guests would donate their time to give talks and host activities.

New branches of Peeka had also been added since 2009. They adhere to the earliest design strategy of being close to the target population, public transportation, and large-volume foot traffic. The new branches also target different citizen groups in the city: the oldest branch, Wanliu, is in the northwest part of the city where there are many universities; the one in Sanyuanqiao is near the city center, where many foreign families live; Jinsong is located among the "move-back" Beijingers, mostly blue-collar workers who returned to the city after their old homes were placed by new apartment buildings; and the Wangjing branch is in a middle-class neighborhood. Annual membership costs 800 yuan, the equivalent of US $120. This membership was designed to be affordable even to those in the lowest income bracket of 30,000–60,000 yuan annually while taking into account the fact that most families are still double-income families.[10]

New people continue to be attracted to the idea of Peeka. For example, I met Meggie, a close friend of the Peeka founders and a Tsinghua alumna whose career path is very similar to that of the founders. When she returned from the East Coast with her daughter and learned of Peeka, she told me she was so thrilled with the project that she started to block out Friday nights from work and family so that she could teach an English class there. For her completely free labor and enthusiasm she was made a Peeka advisor. "My class is billed as phonics, but we designed it in such a way that it includes everything students need to know to learn to read and keep reading," Meggie explained to me as she approached her fifth year of teaching. "This is the reason why the class takes three semesters. I think it's very important to help local kids learn to read English on their own." Her efforts were rewarded, she said, when one mother told her, "My daughter started to get 'A's' when she started taking your English class, and her grade has never dipped!"[11]

For such a library to support all these activities and new locations, the division of labor at Peeka remains pretty informal. As Ming summarized in her breezy way, "We came together and things simply fell into place all on their own. We do what we are good at and keep doing it. Peeka thrives with our friendship."

Birong noted that she used to do a lot of chores at the start of Peeka but that "Yi has a vision." For example, Birong said, "Yi reminded me that I should give the library department heads room to build their authority among the workers." And more importantly, "She sees the best in every person." Birong recalled the many times they interviewed prospective librarians, were uncertain about a particular interviewee, and Yi would say, "Let's see what we like about her, and then we'll see if we can use her strengths." Birong says they all look to her calm presence when problems arise. "Let's wait a minute, and look at this," Yi would say. And they would look at "this" and the problem would get solved.

As the last one to join the group, Aidong found her niche by designing and writing the curriculum for classes of all levels. And Ming says, "I like to make friends and I don't mind dealing with people." So she connects Peeka with outsiders: getting licenses, dealing with officials, and even dealing with the media. She is the spunky one among the founders; her eyes dance with sparks, her hair bounces over her shoulders, and her outfit usually cries out with bright colors, an extension of her personality. Her size is similar to that of her friend Yi, yet her outgoing personality is in perfect contrast to Yi's gentle manners. "I used to tell my advisor that my career was to make friends and pursue happiness. The advisor told me that this isn't a career goal but a personal goal." Now, says Ming, "I can show him that the two goals can be the same!"

The passion and personal chemistry among the founders is infectious. The chief librarian is Cheng Xin, a former English teacher who stumbled into Peeka and offered to quit her previous job and work for them. Many younger staff members and volunteers have also come, seemingly by happenstance, to help at the libraries. People offer to give free talks and lead activities, and prospective parents who come to check out the libraries sign up their unborn children as members-in-the-womb. Many petitions to set up new sites and to take over failing libraries have come in to Peeka's founders, but they remain committed to steady development at their own pace so as not to dilute the quality. They focus on what they already have and aim to make these libraries a true stronghold of their approach to teaching and learning.

News media have publicized Peeka with headlines such as "Tsinghua Graduates Set up Children's Library" and "Let Children Experience American Style Library Fun." As word started to spread, the blogosphere exploded with rave reviews from grateful parents.[12] In 2012, the government finally granted Peeka a license for non-profit organization status. The officer who came to inspect the premises encouraged Peeka to apply for government grants for projects such as helping the city's migrant workers' children and doing more for low-income groups. But funding for these small projects is often limited, government regulations formidable, and, as Ming says, "we're heading in that direction anyway." But they are definitely pleased with the fact that after six years of operation the government has granted Peeka the NGO status that declares the operation worth supporting.[13]

On a cold winter day when I visited the Wanliu branch of Peeka, a fashionably dressed young mother in shiny boots was showing her two-year-old daughter a pop-up book. She had driven an hour to this library so her daughter could spend a couple of hours with other only children in a good environment. She said she hopes Peeka will set up a branch near her: "Their collection is peerless and saves parents the time and trouble of finding good books, especially as we are less familiar with American children's literature." She then puts a large jacket on her daughter and she and the little walking coat of pink head toward the door.

Child-centered Western children's literature finds ready followers among today's young parents and their young children, products of the single-child policy instituted as part of the Reform at the end of the 1970s and early 1980s.[14] These young parents and their children have become a social group more child-centered than the traditional hierarchical society, and they are eager consumers of Western culture. Now, with their expanding opportunities and accumulated wealth, many of these consumers try to become participants in Western culture. They want their children to attend schools in the West in order to escape the intense pressure in Chinese schools and become more international. For them, Peeka is a good place to start.

Reintegration

If Tsinghua and Beida taught them how to become scholars, and America exposed them to the world of children's literature and a different way of child rearing, no one taught the founders of Peeka how to set up libraries in a city they have chosen to call home.

Bureaucracy was the first thing they had to deal with head-on. "At the beginning," Ming recalls, "we learned in the media that the government encourages setting up private libraries. But when we went to a government office to apply for registration as an NGO, no one working there had heard of setting up a non-profit private library and no one knew where to start. They sent us on a wild goose chase from office to office. We then had to look for the exact document on our own and bring it to the registration office. They studied and checked and discussed, and went through a whole rigmarole of making up the process as we went along."[15]

A sticking point in the initial registration was how to categorize the library. If it was privately run, the registration office decided it had to be a commercial enterprise and thus pay commercial tax. "No, no, no," the women replied, "it will be a non-profit organization." But there was no rule to follow for this kind of library, and the mothers could not come up with rules themselves. They paid commercial tax until 2012, when their NGO tax status was finally granted. By that time, Peeka's finances had improved, finally bringing them closer to breaking even.

Could they have made some things easier for themselves? Perhaps. Ming recalls one time when she had to renew the lease for their first branch soon after returning from America. Worried about the legal consequences if the lease expired, she tried several times and was ignored each time. She returned for one last try during the Mid-autumn Festival, bringing along a large box of high quality moon cakes as an offering. "Miraculously, things started to happen," she said. "It may have been coincidence, I really don't know, but the boss just happened to walk by me." Ming still looked amazed as she recounted the event years later. "The secretary introduced us, and he simply signed the new lease." With her sunny disposition, Ming hesitated to brand the event as bribery because the moon cakes were a pittance by the measure of monetary value. She prefers to call her gift an olive branch.

Ming's ultimate frustration stems from the official lack of clarity. "The hardest part of getting things done here is not knowing what piece is missing," she says. "Everything is fuzzy. No one tells you anything, no one wants to leave a trail for possible accusation, and no one wants to take responsibility. There are so many people in the office everywhere that the ball gets kicked around pretty easily. They either don't know, don't care to know, or simply don't care and don't want to know. As the old saying goes, 'one thing fewer is better than one thing more to do.' There is almost an incentive to know less. No one tells you rules or guidelines, so you have to learn the ropes yourself. Moreover," she adds, "you have to educate *them*. Then *they* decide whether they'll help you or not."

The women readily laugh off their own inadequacies, too. On Peeka's opening day, the four of them found themselves the only ones in their sparkly new library on the second floor of the shopping mall. They had not planned an opening ceremony, had not sent out advertisements or invitations, and had not used the firecrackers or celebratory flowers that typically bring attention to a new enterprise. People stopped by out of curiosity and asked, "Is this a bookstore?" "Is this a daycare?" When getting negative answers, they seemed confused and unable to assign this new place to a familiar category. Their first real customer turned out to be an American mom, Tricia, who was walking by with her six-year-old daughter and younger twin boys. She was so thrilled to find a perfect place for her children that she bought a membership right away and became a story hour volunteer until her return to America the following year. "We didn't even think of having an opening ceremony, but somehow the word got out anyway!" The women still marvel at their own naiveté about publicizing this event.

And little did they know that their library would become a focal point of their own deep reflections on their even deeper drive to commit to such a project.

On one smoggy day in Beijing, Ming called and changed our walk to a meeting in her home instead. The PM2.5 count in the air was above 500.[16] "On days like this," Ming said as I entered the spacious living room of her modest sixth-floor condo, "my first thought is: Let's go back to America! I really don't know how to deal with this air anymore! Come spring, there will be sandstorms, too, and everyone will cover their face. I do so miss

California's blue sky." As she speaks, her cell phone buzzes and Ming reads a text. "Look," she says, "here's someone who is better than me. My friend Xiadong just texted me: 'Is there something we could do?' Xiadong always finds meaning in helping others. Since her return from America, she has volunteered at Peeka from time to time, and has even spent months in an AIDS village as a volunteer. But I have a narrower vision," she continues. "I like to be happy. I believe only when I'm happy can I make people around me happy. So Peeka fits my personality. To me, it is about more than helping kids learn. It is where our closest friends on the team face challenges together, forging deeper friendships. Peeka has become a platform where we learn about ourselves and learn how to reach out and help the broader society."

Life in Beijing has become very different from their college days in the city. They have made different choices in their legal status, as China does not allow dual citizenship. With an American passport or an American green card, they need to leave China at least once each year to validate their legal status in both countries. So they take their children to their old homes in Silicon Valley to attend summer camps, seizing the opportunity to recharge in the American culture, do fundraising, and, in Aidong's case, conduct library research to improve her plans for Peeka. America is practically their summer home. For the rest of the year, Beijing is where their families live.

Ming's challenge in raising her curious and intelligent sixth graders at a Tsinghua-affiliated international school is to help them bridge their home in China and their summers in California. "I want them to have a positive view, I want them to have problem-solving skills, but, more importantly, I don't want them to be cynical; I want them to become good international citizens," she states. But teaching her children the right values has been a challenge. Even the simplest everyday thing such as crossing the road, for example, puts Ming in a quandary. She tells her children to follow traffic rules. But soon she realized that if her twins were to follow exactly what she told them they would never get anywhere on their own. There may be traffic lights, yet the streams of bikes and motorcycles never stop for anything, not even children. In the end, she not only had to revise her own instructions, she had to accompany the twins to school every day, basically showing them in person how to violate the rules in order to survive and get things done.

Challenges also come from the TV news. When the children asked their mother why no one had helped an old woman felled by a passing car, Ming was stuck for a proper answer.[17] Should she tell the children that it's not infrequent that a rescuer is accused of being the culprit, and that this was the reason why many don't dare to help? It pained her to see this kind of report, and she herself would definitely help, but did she want her own children to help on such an occasion? She is still not sure. "To teach children the right values, I feel I need a bigger support system from society, and the children need a more confident society within which to learn trust. They're still too young," she concluded.

To Ming, this is where Peeka comes in. "It may be small, but in the long run it will create a positive impact on society, especially if we start them young." She quotes the old Chinese saying—as do other cofounders—that it takes ten years to grow a tree, but a hundred years to grow a man. "We want to put sparks in children's life," she says. "We can't reverse social trends overnight, but we can create positive, long-term influences." Ming knows the importance of starting young, perhaps from her mother. She grew up in the farthest reaches of the nation in a farming community near Aksu—closer to Kyrgyzstan and Kazakhstan than to any major Han Chinese cities. In the early years of the new China in the 1950s, her parents were sent to work in the Xinjiang Production and Construction Corps as part of the government's effort to open up the west. With few educational opportunities there, her mother took Ming and her older sister in their early teens on a long trip to find a good school she had heard about on the radio. The mother and her two girls forded swelling seasonal rivers and hitchhiked on trucks and tractors all the way to that school in Aksu, the largest nearby town. Once there, she persuaded the school to keep the girls as its first boarders, a practice no one there had heard of, before she returned to her work unit. In 1979, the year Ming took the college entrance exam, the entirety of Xinjiang province—an area of 1.6 million square kilometers of which only four percent is habitable and with a population presumably smaller than today's approximately 21 million—was given only ten slots for students to go to Tsinghua. Ming clinched one of them.

For Aidong, the challenge of finding positive forces while raising a child in China hit home when she learned that her daughter had concentration problems at school. At first she was dismissive: Both she and her husband had sailed to the top of the same system with ease, so how bad could it be for their children? She decided their daughter could use some discipline from a Chinese school, but the test-oriented education at the expense of attention to an individual student's needs in classes averaging 40-50 students caused her daughter to crack one day, and she adamantly refused to go back to school.

On a cold winter day in 2013, Aidong was pensive as we sat at a café right outside of Tsinghua University that she used to frequent as a student. Her daughter's pain and struggle weighed heavily on her. Aidong has a tall and athletic frame, and her darker complexion (unfashionable in China) would fit perfectly under the California sun. Running her hand through the hair that tapered off at her shoulder, she asked with a sigh, "How can I make it up to her?" She wished she had been on her daughter's side all that time, instead of siding with the school to discipline her.

Having lived a life defined by academic excellence and career success with relative ease, Aidong for the first time realized the challenges of being a mother as she faced her daughter's problems. She transferred her daughter to a private school and started attending classes and talks on child-rearing with her customary seriousness and dedication.

While reflecting on the message of a recent talk—that a happy family is based on love, but an unhappy family often focuses on performance—she compared her own upbringing to that of her daughter. Her own parents had been swept up in the Cultural Revolution during her school years, but had always been proud of her stellar performance at school and came to expect excellence from her. So when her mother learned that Aidong had returned from the States to co-found a children's library while being a stay-at-home mom, she stopped bragging about this daughter to her friends. But this changed when Peeka began to make a buzz in the media. "So often parents wear their children's success as their own glory," Aidong sighed. But with a well-trained scientific mind that sees both sides of issues, she saw in her daughter's ordeal a chance to understand her own role as a daughter and a mother: "Only when a women becomes a mother herself can she truly start to understand her own parents," she notes. Parents only expect the best from their children, and this expectation is often conditioned, as well as limited, by the time and the society that have shaped them. This awareness brought her closer to her parents. From that point on, Aidong could see that her parents had been limited by their circumstances and lack of options, and that she now has the advantages of education, financial security, and the ability to explore different worlds with ease. Today, with her greatly expanded awareness and solid financial footing, she says she feels an obligation to do right by her own children.[18]

Stand by Me

"Peeka is our baby," the women like to say. Even with their academic and professional achievements, they readily acknowledge that none of them would have been able to do this alone: "Not one of us would be able to give 100 percent, and it certainly takes more than that to run this." At the beginning each partner put in more than their fair share of 50 percent as they ran around to get things started. Ever since Birong returned to the States in 2007, the other three have kept things running and have worked to reduce the deficit. But all three insist, "If we're the mothers raising this baby, Birong is the birth mother."[19]

Birong now visits Beijing about once a season. Her visits are billed as the "Bi'r Festival." After missing her a couple of times during my year in Beijing in 2012–2013, we finally met at the Spring Bi'r Festival. "I want to make this a bright and warm part of someone's childhood memory," Birong says of her dreams for Peeka. "I feel so fortunate that my friends are working with me to make this dream a reality even during my physical absence." All four women are in their mid-forties now. Birong even shares a similar physique with Yi and Ming: she is of medium height, has a trim figure, and her shoulder-length hair frames a gentle round face. Birong looks very fair sitting by the sunny window of the café; even her eyes are a shade of light hazel. Recalling the moment over seven years ago when she decided on the spot to rent the first library space, an almost childish delight lights up her face. "The group of us was to leave on a scheduled trip in just a couple of months, and I wasn't sure if we should be starting something new at that moment. The realtor urged me to take it before the rent went up, and I made an anxious call to Luo Ming. Guess what she said? She said, 'If we can't do it in two months (meaning before the trip), we won't do it in two years.' So without hesitation I rented the space." As it turned out, they were duped into renting the place just before all the businesses around it went bust, forcing them to move to a new location not long thereafter.

When she moved back to California for family reasons, she felt guilty for leaving a not-yet-two-year-old Peeka to her friends. The best she could do while away, she decided, was to ensure the financial well-being of the library. To her, Peeka was and to a certain extent still is a lonely outpost of a purer vision of childhood learning where immersion in books allows children to find a world of their own.

Birong comes from a professor's family, as does Yi, and all four women achieved their academic excellence for the love of learning. This is the gift they want to share with children in the sunlit rooms of Peeka.

Yet in the increasingly commercialized world of today, even holding up a vision comes with a price tag. "Sometimes it's hard to impose our ideals on the people working for us," says Birong, who is in close touch with the day-to-day operations of the library. "People need to make a living and we need to take care of them in order to keep the library a happy place." Birong feels fortunate that she and her husband have cashed out on hi-tech stocks and are on a good financial footing. "I don't care for fancy handbags. I feel that my best contribution to Peeka is to give it what we call a 'blood infusion' to keep it out of the red and financially viable. And helping Peeka has helped me learn how money can help bring about happiness and pride that's profound and lasting." Birong explains that she has found ways to finance and sustain Peeka from the beginning through a "donor advisory fund" in different foundations where they are donors. Through these foundations, Birong is able to legally channel money to Peeka libraries. Other founders also help hold fund-raising events on their summer trips to the States, in addition to their expanding fee-based offerings in the libraries. Peeka's net annual loss had been around 20% of the total operational cost, and this has been halved in the last two years.

These women have acknowledged that many practical things could have been handled better, particularly had they not been blinded by their own idealistic dreams for the library at its inception. Some oft-quoted examples include: being talked into renting a piece of real estate that would soon lose its value instead of purchasing a more permanent site for the library before the city's financial boom, being too conservative about expanding, and not having pushed for a non-profit tax status at the very start. But with Birong shouldering a big part of Peeka's finances, the women are grateful they don't have to chase after financial opportunities at the cost of keeping true to their original goal of making learning a joy.

It's the Tradition

As full-time mothers, the women are now able to see in a very personal way that it was quality education that enabled them to get into Tsinghua and Peking (Beida) University; and that from there they were able to move into the greater world with a vision and awareness of life options and possibilities. They may not have realized it at the time, but when they entered Tsinghua and Bei-da, they became part of a tradition.

Tsinghua University was founded in 1912 as a western-style

preparatory school preparing students for studying in the US. It was founded with part of the money from the Boxer Indemnity paid by China to the United States. Early in the Republic Era of the early twentieth century, many beneficiaries of the Boxer Indemnity Scholarship program who were impressed by the advanced technology and economic prosperity in the US returned home to advocate "saving China by science and technology." One of its most famous alumni, Hu Shih, introduced Columbia University professor John Dewey's famous theory of pragmatism to China, vehemently opposed feudal ethics, advocated for individual liberty, and promoted ideas such as democracy and science. Founded in 1898, Beida is the nation's first modern university. It has been a center for progressive thought since the New Culture Movement in the 1920s. Many of the most prominent Chinese leaders and thinkers went through its gates: Mao Zedong, Lu Xun, and Hu Shi, for example. Standing side by side in the nation's capitol, these two top universities have been the hotbed of reform movements in China's modern history. Generations of their graduates continue to return from America with fresh vision and new skills as they become leaders in all fields. This traditional trajectory of a Tsinghua or a Beida student has delivered these women to a stage in life where they have come to own their mission of introducing to China positive aspects of US education and building a bridge to the wider world. As the Peeka motto says, "Peek-a-book, Peek a world."

Peeka's plate was full at its year-end meeting a year after Tsinghua's centennial. To deal with the over-protective parents and grandparents of only children, the founders drew on their connections both in China and abroad to plan new activities for both parents and children in the coming year. They added parent education lectures and activities that lure parents away from monitoring their children during classes or activities, and took downhill skiing out of their camp program to ease parental concerns. They planned to reach out to more communities in and around the city, donate more books to schools of need, and send out mobile libraries to underserved communities, especially where migrant workers congregate. At the end of the meeting, they decided to accept a recent request to establish a franchise in Shanghai, an expansive step for a very prudent group of mothers. No grand views, just peek a book at a time—one child, one community at a time—as Peeka grows into another year.

Chapter 4 Notes

1. Reports and postings can be found in all major Chinese news media, including *China Daily*. Young parents who return to China for a short stretch of time—on vacation or for a short job assignment—have found Peeka to be a good place for their children, and enthusiastically recommend Peeka to their fellow parents in online forums for overseas Chinese communities. These children growing up outside of China either can't read Chinese, or can't read fluently on their own. So Peeka becomes a good transitional, educational venue to help them ease into the Chinese-speaking environment. For example, a mother begins her enthusiastic note on the *London Chinese* site as follows: "This summer when you take your children back to China for their long vacation, have you thought of how they'd spend it well? Peekabook's "Root-Searching Journey" [a summer camp program] will be a good choice for you" (April 14, 2010). It then explains in great detail Peeka's programs, and even lists schedules for reference (see http://www.londonchinese.com/forum. php?mod=viewthread&tid=17904).

A *Workers' Daily* article compares various privately owned libraries in the capital city of Beijing and says of the Peekabook House, "no doubt a pathblazer in building private libraries" ("Private Libraries: Fraught with Frustrations in Growth 私人，成长多恼," *Workers' Daily* 工人日报, http://acftu.people. com.cn/n/2013/1216/c67502-23845698.html (December 16, 2013 [accessed March 29, 2016]).

2. The founders strive to have the best editions of all their English books by purchasing them from authorized publishers and distributors. They particularly pay attention to the young children's collection, as children of that age have more contact with the books. This collection in particular has always been a big draw. To save on postage, "Many of these books have been carried back from abroad on our backs." Due to their popularity, these books are heavily used and need frequent replacement.

3. The president of the Capital Library Ni Xiaojian's full quote in the article is as follows: "The Capital Library has been paying attention to [the rise of] private children's libraries and is working on how to support them. For those private libraries that are completely devoted to public welfare and with fixed locations, we can show our support by donating some books. We can also send over some professional library managerial staff to help them raise the level of children's reading abilities." Ming Zhong, ed, "The Capital City Blossoms with Private Children's Libraries, Games Cultivate Reading Habit 京城兴起私人儿童，游戏培." *Cultural China* http://culture.china.

com.cn/book/2010-08/23/content_20766338.htm (August 23, 2010 [accessed March 28, 2016]). Also see Lijin He何利進 and Dong Chao 董超, "Returnee Libraries Start to Increase 海归多起了" *People's Daily Overseas Edition* (May 31, 2013).

4. The guidelines for the Capital Library have a cut-off age for children: Section A in the building takes everyone above 13 years of age, and the children's section takes everyone below 18 years of age. But clearly there are other rules not mentioned in the official guidelines. When a mother put up a comprehensive list of libraries around the city of Beijing that accommodate young children, she wrote of the Capital Library: "students under 16 can check out for free, children under 8 should pay 80 yuan deposit in addition to the library card, those above 8 pay 30 yuan deposit in addition to the card." But the rules and regulations of the Capital Library vary from site to site, and are subject to change over time. So this information can only be taken as a general reference.

5. Although this sentiment has never been clearly spelled out, one can easily sense it in every piece of reporting on the library. The academic credentials of these founders always have a prominent mention in articles.

6. This number was given to me by these Tsinghua graduates themselves.

7. During the years 2012–2013, I lived with my family on the campus of Tsinghua University. Due to the proximity of our homes, I had more interaction with Ming and Wang Yi. Both are members of a much larger group of female returnees in the city, and, in general, this tight-knit group of women had frequent activities of outings, concerts, shows, lectures, and interviews, along with simple get-togethers for a birthday or a return visit of one of their friends. Aside from activities with this group, I also had repeated in-depth interviews with them separately. This interview with Ming was one of the few conducted in her home.

8. Feng told me this when I asked him why he returned.

9. These returnee moms take full advantage of their circle of friends both inside and outside of the country, who are usually successful professionals. So their talks often draw a large crowd of parents. I have attended some of these talks. To attract more young professionals, the talks are held in the evening, and they usually pack the largest activity room in the library. When a talk is over, the organizers generally take the guest out—usually their old friend Beijing for a visit or work trip—and have a cheerful get-together. Restaurants are usually a good venue for social gatherings since most people living in the city do not have homes large enough for such events. Many of these returnees have larger homes in addition to the apartments near their children's schools, but the larger homes are usually located on the outskirts of the city and are too far from the city for a regular commute.

10. The official regulation in 2012 for the lowest salary in Beijing was adjusted from the previous "no lower than 1,260 yuan per month" to the new number of "no lower than 1,400 yuan." A chart from the official newspaper *Guangming Daily* 光明日报 was posted on an online site of *Xinhua Net*, ranking the average annual personal income for the year 2014. It listed Beijing's as 43,910 yuan, the second highest in 31 cities and provinces surveyed,

an 8.9% rise from the previous year. ("Announcement of Personal Income Ranking in 31 provinces: Shanghai First and Beijing Second 31 个省份人均收入排行公布：上海最高北京第二," *Xinhua Net,* http://news.xinhuanet. com/fortune/2015-02/27/c_1114459674.htm (February 27, 2015 [accessed March 28, 2016].)

11. Meggie was a Tsinghua classmate of the three founders and has traveled a very similar path to theirs.

12. Some of the major publications that have reported on the Peekabook story include: *People's Daily, Reference News, Beijing Daily, China Development Brief,* and *Chinese Women.* The blogosphere has carried real-time reports and comments on Peeka and its activities. At the same time, Peeka's staff is also diligent and skillful in publicizing their activities. The staff is usually made up of well-educated, tech-savvy young women. Peeka also recruits college students as temporary volunteers. Their successful publicity campaign is the result of the concerted effort of a talented media team.

13. In 2011, the head of the Ministry of Civil Affairs, Li Liguo, announced at a meeting that three types of social organizations could directly register with the government as independent social organizations, and that they would no longer have to follow the earlier rules that demanded a sponsoring government organization. These three types are: charity, social welfare, and social services. Peeka was able to take advantage of this new regulation at the time as an organization for social services. See Jinhui Guo 郭晋, "Three Types of NGO Registration Requirements Loosened, Stepping Up Reform in Management System 三类NGO注册条件放，管理体制改革起步." This report on *No. 1 Financial Daily* 第一经日报 can be found at *Sina,* http://green.sina.com.cn/2011-07-13/135322806932.shtml (July 13, 2011).

14. For the country's "one-child policy," please see note 15 in the Introduction.

15. Ming mentioned that this was a notice in a local newspaper and that she no longer has that first document. But the registration for an NGO has been notoriously complicated and frustrating, primarily because it demands a government organization as a sponsor. Yet no government agency was willing to take on an unknown entity.

16. "PM" stands for "particulate matter" and a PM2.5 count is a common way of measuring air quality. According the US EPA, the US national average PM2.5 Index between 2000–2014 about 12. Data last retrieved in February 2016 at: http://www3.epa.gov/airtrends/pm.html.

17. One landmark event that shocked the Chinese people into deep reflection on an individual's social responsibility was regarding the little girl called Yueyue. She was hit by a car and left unattended on the street for hours. A detailed story of Yueyue and its repercussions in Chinese society in general can be found in Evan Osnos, "Passing By," in *Age of Ambition: Chasing Fortune, Truth, and Faith* (New York: Farrar, Straus and Giroux, 2014), chapter 20. This accident happened in October 2011 and was still fresh on people's minds at the time of our talk.

18. The family finances of these women are generally secure, as their

husbands are usually founders of very successful start-ups and now companies. Some of them are founders of more than one company.

19. An update on these founders' whereabouts: In late 2013, Aidong returned to Silicon Valley with her children. Ming returned to work at Stanford in 2014, also with her twins. A combination of factors has drawn these mothers back to the States: As the children are growing older, the mothers are positioning them to get into good American universities; the pollution in Beijing, as well as other concerns mentioned earlier in the chapter, food safety included, all pushed these mothers to return to America for the final years of their children's high school. It has never been easy for any of them to live either in America or in China, as each place offers something unique that they deeply cherish. In our talks they constantly yearned for "a place where everything we like can be all put together."

CHAPTER 5

"We Are Not Witnessing History, We Are Making History": Chen Datong and Shanzhai Cell Phones

The One Percent

On a hot and muggy summer afternoon I sat in a classroom at the Golden Seed, one of Beijing's three government-sponsored, hi-tech start-up incubators.[1] Dozens of aspiring entrepreneurs filled the room, awaiting the arrival of Chen Datong, co-founder of OmniVision and Spreadtrum and CEO of WestSummit Capital. Next to me sat Liu Yun, a recent Tsinghua graduate. Liu told me he has a video game start-up company and that, in Chen's stories, he finds validation for his team's recent decision to expel one of the founders. "We weren't sure if it was the right thing to do, and we were delighted to hear Chief Chen had to do the same in his company!" he explained.

The crowd suddenly becomes hushed and sits up straight as Chen Datong enters. His bearing is much the same as it was when we met the previous week:[2] quietly confident and perfectly at ease. At age 58, he is lean and his eyes seem to recede into perpetually deep thought behind thick glasses. He sits down, loosens his shirt collar, and adjusts his dark jacket, and waits for the sponsor's effusive introduction to finish. Then, amidst a round of applause, he takes the lectern and surveys the audience.

When he begins to speak, everyone falls silent. Instead of taking notes, they snap photos of Chen's outline on the screen with their cell phones. "You never get this kind of lecture in college," Liu told me in an excited, hushed voice. And then Liu is completely immersed in the talk, hoping to hear more advice to guide his nascent start-up.

Chen is one of the most well-known returnees in China's hi-tech world today. Revered by many as a godfather of the IT industry in China, Chen is a poster child of China's economic miracle. A revolutionary youth turned capitalist, he co-founded companies and turned them into gold. In 1995, he co-founded his first company, OmniVision in Silicon Valley, serving as its Vice President of Technology. Under his leadership, Omni-Vision's single chip CMOS image sensor revolutionized camera imaging technology, enabling the company to become one of Silicon Valley's most successful companies.[3]

Chen returned to China after OmniVision went public in 2000, and co-founded the Shanghai-based company Spreadtrum, serving as its Chief Technology Officer (CTO). He led his team to produce the first single-chip dual mode TD-SCDMA/GSM baseband chip, revolutionizing the way cell phones were made and used. With this multi-function single chip, Spreadtrum challenged the domination of the Chinese market by Western companies such as Nokia and Motorola by creating overnight a new and massive bottom stratum of low-income buyers such as migrant workers. Spreadtrum's IPO in 2007 raised funds of just a little over US $100 million. Six years later, the Chinese government-owned Tsinghua Holdings acquired it for $1.78 billion.

During his lecture, Chen declares that "The global market has undergone a historical transformation." He elaborates: "This is the first time since the Industrial Revolution that a new market has emerged to compete with North America. This market is China, leading a pack of new emerging countries." With some pride, he notes, "At a speed of development two to three times faster than the American and European markets, China has disrupted the global game, and is creating a new order." As an illustration, he dissects the growth of Spreadtrum. "The disrupted global order created disorder, as well as opportunities to try out new technologies and new business models."

Chen enumerates the rapid changes in China that are familiar to his young audience, and then makes a sharp turn in his argument. On the screen, he displays the following slide: "The One Percent Rule: the 1% true market opportunity is seen only by 1% of the people."[4] He explained that 99% of changes prove to be ephemeral and only 1% of enterprises have staying power.[5] Assuming everyone in his audience would like their start-ups to be in that one percent, Chen focuses on the importance of leaders and leadership that are, he says, the keys to a business's staying power. Time after time and case after case, Chen returns to the example of Spreadtrum to illustrate the importance of leadership vision and the ability to lead when there is no easy model to follow. Then, his PowerPoint drives his message home: "The controller of the market has the power to make the rules."

With Spreadtrum's success in the low-end phone market, Chen not only makes the rules, he also packages the Spreadtrum story and defines it in his own terms. In the very inception of Spreadtrum, for example, he sees perfect parallels to the birth of the Chinese Communist Party (CCP). At the start of 2001, Spreadtrum began with a team gathering at the Silicon Valley home of Qiao Peng, one of the planners. Once the team decided on the name Spreadtrum, he explains, they registered the company in the Cayman Islands.[6] This beginning of Spreadtrum might appear to be a traditional American start-up story in a typical Silicon Valley garage, but it is differently framed by Chen.

In one of the lecture notes he emailed to me, he compares the start of Spreadtrum to the birth of the Communist Party in 1921, and the CCP's founding of the People's Republic of China (PRC) in 1949. Speaking of the CCP, he writes: "Almost a century ago, over ten people gathered on a boat in Zhejiang. They brainstormed for many days and nights, drafted a manifesto, assigned some positions, and a small 'company' quietly came into being . . ." Describing the similarly modest founding of the CCP, he finishes thus: "They eventually surrounded the cities from the countryside, and in 1949, the 'company' successfully went public, with the IPO code 'PRC.' This 'company' has always led in one number: it has the largest number of employees, making a great contribution to global employment!"[7]

Chen's China Story

Chen is now based in WestSummit, his latest company. The office of this venture capital company is in the heart of Beijing's financial center. Up on the ninth floor in Vanton Tower, the office suite is still humming at 5:30 in the evening when I arrive for our first meeting. Chen's video-conference is running overtime, the secretary informs me apologetically. Minutes into my wait, Chen gently knocks on the door of the waiting room, apologizes, and then vanishes.

In Chen's brief absence I am left to myself in an intimately sized waiting room. A set of black leather couches faces a blank wall against which stands a large TV screen on a stand plus a large writing pad on an easel. Two sprays of moth orchids on stands separate the chairs from the couches and add a soft touch to the room. The large glass cabinet tucked in the corner is empty except for a couple of bottles of whiskey. Under the couch is a pair of dumbbells. Through the glass wall on one side, I see the congested evening traffic weaving in and out of the office towers on the streets below.

I came to meet Chen through Ming, who was his student at Tsinghua. The two have become close friends over the years since her graduation. Speaking of Tsinghua, Chen recalls that he wanted to be an "Einstein" when in college. But his time in America, he says, turned him into an entrepreneur. Now he thinks this latter role suits him better. In the quarter century or so since he left the university, he has moved a long way from his scientist dream to become a godfather of hi-tech entrepreneurs. As he adjusts his life goals and recreates himself along the way, he says he finds his leadership role most exhilarating.

The responsibility of being a leader, Chen tells me, "gives a person a chance to find his own potential. At the beginning, I thought I could be a fairly good engineer. But now I know I could do more, I have other potentials. How do I know? When new demands arise, the leader has to fill in and do the job before he can find an appropriate person to hold that position. As a result, the leader learns many different jobs he didn't know he could do." But more importantly, being in the leadership position has afforded him a chance to learn lessons he had never learned in school.

"You always have to be an optimist," he says. "When people tell you that something cannot be done, you just have to try it. All who succeed in life are optimists. I found this during my years in the countryside, and I see this in the business world." From these two worlds outside of school, he has learned that the real world is not black and white like the science he studied in school. "There are no absolute good guys or bad guys, just as I have my own strengths and weaknesses." As he says with much conviction: "Therefore, you don't hold others to improbably high standards."

Chen is constantly summarizing lessons learned. His leadership position offers him a higher perch for a better understanding of people, but more importantly, it helps him make connections to the leadership he has grown up with: the Chinese Communist Party. In invoking the CCP campaign for a new China, Chen sees his own endeavor as part of the tradition to make China great again.

This point is amply elaborated with the case of Spreadtrum in another one of our interviews in that same room. It is the end of another long workday for him, but Chen seems rejuvenated by telling his Spreadtrum story. Stepping out to remind the young lady at the reception desk to fill my teacup with more tea and bring one for himself, he plunks down on the couch and relaxes in his loosely-fit, tieless ensemble of a dark-colored jacket and pants. Although Chen is no longer involved in the day-to-day operations of Spreadtrum, he sits on its Advisory Board of Directors. Spreadtrum was, and would always be, he says, his China story. Skipping over his decade in the States and the story of taking OmniVision public, Chen eagerly focuses on 2000, the year he returned to China from Silicon Valley. His only comment on the move is, "Silicon Valley was getting crowded."

What he found back home was an untapped market: a dozen of the country's major, state-owned semiconductor companies had collapsed under the onslaught of major international companies' entries into the Chinese market in the late 1990s. Huahong, the company established in China with much fanfare in 1995, was struggling under market pressure and the handful of integrated circuit (IC) design companies were making only simple products such as phone and transit cards. At the time, blazing a path to technological success in China seemed like a difficult dream to accomplish.[8]

But Chen was encouraged by the fact that the government was clearly trying to change the situation. In the fall of 2000, the government's Information Department issued document #18 and set up Zhangjiang Science Park in Shanghai in an effort to attract and facilitate the development of the software and IC industry. Responding to the government's encouragement, some of Chen's friends and students returned from Silicon Valley to found start-ups. Chen was intrigued, but took his time to find a niche that would make the most impact.

The wait ended after a talk with the vice minister of the National Information Department. Minister Qu was candid about China's sorry state of affairs in the cell phone market: All the chips were imported from abroad. To break the foreign monopoly on the Chinese cell phone market, Qu's department had been pouring money into fostering China's own since 1997, yet nothing had come out of that effort. At the time they spoke, Vice Minister Qu had given up on 2G technology. But this was the moment Chen found his chance to lead China into the next generation of cell phones. Two eager parties soon formed a partnership: the government promised initial funding of US $30 million to start a telecommunications company, and Chen would assemble a team.

After a few phone calls to Silicon Valley, Chen quickly assembled a formidable leadership team: Qiao Peng, an experienced CEO in the field (founder of two hi-tech companies by that time) was first invited to Beijing to give advice; Wu Ping, the director of R&D at MobileLink at the time, and someone who had already been planning to start his own company, became an instant partner and co-founder. The two soon recruited three more co-founders to match their own credentials: Ji Jin, a Tsinghua alumnus who had received his MS and PhD from the University of Michigan, and later worked at IBM's Watson Institute as a member of the team designing CPUs; Fan Renyong, a graduate of Nanjing University who had earned two master's degrees and a PhD from Michigan State University, was a co-founder of Luxxon company and later worked as chief engineer and design manager at Sun Microsystems Trident; and Zhang Xiang, who had graduated from Zhejiang University with both an MBA and an MS in communications engineering and was at the time working on software development and product management in the greater San Diego area.

They decided to set up their company at two locations simultaneously, an unusual move at the time: there was an IC design team in Silicon Valley that took advantage of the mature technology there; and a software team in Shanghai's Zhangjiang Technology Park that took advantage of a vast talent pool and government incentives.[9] "We did this because we knew we were facing a global challenge," Chen explains. Within months, they had put together nearly 100 people in both locations combined, with the Chinese team as the main force. One fifth of the Chinese team was made up of overseas returnees, while the rest were the best-trained young college graduates in China.

Class of '77

The core members of these earliest teams consisted of the earliest batch of Chinese students who studied abroad after the Cultural Revolution. Chen himself was a member of the class of 1977 (marked by the entry year in China), the first class admitted to college after the Cultural Revolution. In 1977, China resumed college entrance examinations after suspending them in 1965. Among the 620,000 exam takers, 4.7 percent won admission to the nation's existing colleges.[10] Chen and his classmates who went to Tsinghua University, the nation's top engineering school, were the cream of that precious crop. He graduated a dozen years later with the university's first classes of BS, MS and PhD graduates from the Department of Electrical Engineering.

Chen looks back with a self-deprecating smile, "I thought I was pretty good at making things work. Ever since I was very young, I've been a fanatic about anything wireless." Talk about his earliest inventions still sparks a mischievous twinkle in the CEO's eye. His middle school inventions included a circuit powered by the rear wheel of a bike. He made his own photo enlargement machine, as well as a timer for developing photos. And in the countryside after high school, he fixed everyone's radios and watches and all things mechanical and electrical.

But at Tsinghua University he met his match. One classmate put together what was perhaps China's first black-and-white TV set with foraged parts. When the TV consistently blacked out during the best part of each show, they realized that another one of their classmates had invented a remote control interference device! Chen's own invention with these friends was a coin-operated washing machine that served the entire class.

In retrospect, his graduate school experience at Tsinghua was a dry run for setting up Spreadtrum. In order to set up a class activity fund, the 18 PhD students in the class came up with a business model that "N years later turned out to be rather close to a model resembling that of Silicon Valley's start-ups,"[11] he proudly explains. The commission paid to the one who brought in the work was 10 percent, the reward for actually doing the job was 70 percent, and the remaining 20 percent went to cover class expenditures. Within the year, their combined personal income from a handful of projects added up to a total of 40,000–50,000 yuan. At a time when Chen's own monthly spending money was 56 yuan (about US $9 with today's exchange rate), the personal income for the group was astronomical, and their class had a starting fund of 9,000 yuan, making them and their class the richest in the entire university.

The one project he remembers best is the reverse-engineering of a chip made by Texas Instruments. One of the 18 classmates in Chen's PhD class learned that a factory in Tianjin tried to make duplicates of an imported Canon copy machine. The best engineers hired by the company could not figure out how to reproduce the central circuit board, the "brain" of the machine. Out of desperation, the company gave in to the students' request to give it a try. But instead of showing them a working central circuit board made by Texas Instruments, the factory handed them a discarded one, because they had no faith in the students.

But the factory leaders did not realize the students' abilities. "We had all the major fields covered among us: software, hardware, chips, system engineering," Chen recalls. So they put together a crack team. That summer vacation, the team performed a reverse-engineering of the circuit board by dissecting even the minutest parts under a microscope. When Chen finally finished manually reading and rereading the tens of thousands of bits of the software code for the fifth or sixth time in one month, he almost collapsed from exhaustion. "But nothing beats the reward of the moment of truth when that machine came to life with the board that we had put together!" Chen exclaims.

By the standards of that time, they believe this reverse-engineered circuit board was the most advanced in China. If they set up their own company to commercialize the board instead of handing it over to the Tianjin factory, they calculated the profit margins could make them very rich. But that was 1989, the graduation year of this first class of doctoral students from Tsinghua. The country did not yet have a system in place for a start-up, and in the aftermath of the student movement, the group dispersed upon graduation. Chen jocularly states that he took the route of "foreign exile" to America when he set foot in this foreign land for the first time.

The Magical Chip
Over a decade later, in 2000, China was not only ready but also eager to offer Chen the chance he did not have at graduation.[12] Chen and his team had the minister's support, as well as the government's initial funding of $30 million, and a spot in the government-sponsored science park.

But the first bump came early on, even before Spreadtrum got underway. It turned out that the $30 million initial funding from the Chinese government was to be given out through a first installment of $1.5 million in the first year and another $5 million the second; the remainder would depend on the company's performance. This sent the founders scrambling for more funds. But America had just been hit by the hi-tech bubble bust, making fundraising impossible at that time. At this point, months of intense preparation were at risk of coming to naught. By March 2001, the co-founders of the company agreed that without a meaningful breakthrough within a month they would have to abandon the project. As their plan was about to go under, Wu Ping returned from Taiwan with the money they needed on the table: the CEO of Taiwan's largest semi-conductor design company, Mediatek, had decided to supply the initial funding of US $6.5 million and Spreadtrum was then born in April 2001.[13]

Armed with a top-notch team and Chen's own success with his OmniVision, the group set its sights on creating cutting-edge technology. At the time, their objective was to beat the Japanese to the market with 3G cell phone technology using the more commonly employed WCDMA standard.[14] "Every returnee from Silicon Valley wants to start by going after the cutting-edge technology," he notes, speaking of his own initial vision for Spreadtrum as its CTO.

But the reality in China interfered. It presented the founders of Spreadtrum with two options: If they wanted to demonstrate their technological prowess, they could develop a cutting-edge product using the WCDMA standard and go head-to-head with the forerunners in the field; or they could try to build something completely different with China's own TD-SCDMA standard, breaking the foreign monopoly in China's cell phone market. TD-SCDMA had been proposed by the China Telecommunication Research Institute. Although it had been approved by the International Telecommunication Union in 2000, no one in the field at that time had yet produced devices to support the system. With limited initial funding, the team decided to play the home game and produce the 2.5G technology that would support China's own standard.

More setbacks occurred in this initial stage of the company. Just as the company was about to test its first product, 9/11 hit America. Spreadtrum's second round of funding was delayed with the possibility of vanishing altogether at that critical time of the product's development.[15] What Wu Ping, the company's CEO, did to solve the company's financial crisis has become legendary: To make the remaining money last longer and to allow time to find new money, everyone received a drastic pay cut: employees' pay was cut by 50 percent and the founders received no salary for months. "There were no complaints throughout that period," Chen says with amazement and pride.

The collective belt-tightening paid off. In 2003, the delayed funding of nearly US $20 million arrived, averting the most serious crisis the nascent company had faced. A year later, Spreadtrum fulfilled its own promise of bringing out the first prototype within a year. This was the world's first single baseband chip, TD-SCD-MA/GSM (-SC8800A). More money followed:$35 million arrived for the third round of funding in 2004.[16]

This magical chip combines analog and digital processing devices and a power management device (ABB/DBB/PWIC)[17] into one, thus reducing space and battery consumption without sacrificing functionality. At the same time, it can operate on two different networks (TD-SCDMA/GSM)[18] for voice and data transmission, allowing the same cell phone to function in areas covered by different networks. This simplification makes the chip much smaller and cheaper to manufacture.[19] But more than the technical breakthrough, it also had a national significance. In a market dominated by the WCDMA system, this chip was the first product to support China's own TD-SCDMA system, forcing consumers to choose between these mutually incompatible systems.

The first single GSM[20] chip went into production in 2004 with a run of two million chips, taking three percent of the market in China and attracting much attention. The next year, it was awarded a prize for technical progress, which was the main reason it won the fourth round (pre-IPO) infusion of funding of US $19.5 million in 2006 from most of the funders in the previous round.[21]

The Turnkey Phone that Set Off the Shanzhai Storm

Once the cheaper and smaller 2.5G chip was ready for mass production, Spreadtrum found that native Chinese vendors did not have sophisticated enough devices that were compatible with this more advanced chip. So the company had to develop a "total solution." They decided to design a complete ready-for-market handset with the chip inside. All the vendors had to do was put their brands on, make some cosmetic changes if they wanted, and put in their own entry codes to make each phone their own. Some experts called this a "turnkey solution": just as a car handed over to a customer requires only the turn of a key to start it, the phones handed over to vendors required only minor actions before they were sold. These phones later came to be called *shanzhai* phones, or "mountain stronghold" phones.[22]

They got to the market just in time. "The shanzhai market was just exploding in 2004–05," Chen recalls. Along with the dominant Taiwan company, Mediatek, they blanketed the market with rock-bottom-priced cell phones. As Chen often claims, Spreadtrum was the co-founder of the shanzhai phone market.[23] In other words, only three years after Mediatek helped Spreadtrum get on its feet, Spreadtrum had grown to be its strongest competitor. During a Skype conversation, Yang Zheng, a friend of mine and veteran engineer in Silicon Valley, told me: "The turnkey solution enabled Mediatek to dominate 90 percent of the market until Spreadtrum adopted the TD model."[24] The TD model is China's own standard that Spreadtrum's new chip operated on. It is incompatible with the WCDMA model that Mediatek and all the other companies operated on at that time. Spreadtrum went head-to-head with all the mature companies in the market by blanketing it with this new chip that's far cheaper, longer-lasting, and has all the functions.

"Overnight," Chen recalls with pride, "hundreds of vendors materialized in Shenzhen, reducing the price of a cell phone from 2,000–3,000 yuan (US $400–500) to 400–500 yuan (US $50–90). We transformed the market overnight, and put cell phones in the hands of migrant workers." Continuing his comparison of the CCP with Spreadtrum, Chen points out that their strategy was to "surround the cities from the countryside," a renowned Communist Party strategy. In other words, they combined low prices and profit margins with huge sales volumes to start from the bottom of the market and usurp those at the top.[25]

Shanzhai turnkey cell phones took the market by storm in 2004. The following year, about 40 million sets of shanzhai phones were sold, and by 2009, the number had reached 145 million.[26] With their rock-bottom prices, shanzhai phones created an entire bottom stratum of the market. In 2006, for example, there were 48 million new cell phone users, and the trade-in rate for new phones had increased from 30 percent in 2005 to 40–50 percent. In addition to creating an entirely new section of the market, shanzhai phones also captured market share from both national and international brands, but initially mostly at the expense of domestic brands. The 50 percent market share of the domestic brands—a major milestone reached by domestic markets only in 2003—was reduced to 30 percent in 2005, causing widespread panic.[27] According to Chen, the appearance of shanzhai phones also reduced the market share of the major foreign brands in China (Apple, Samsung, Motorola, Sony, Nokia, and Ericson) from about 68 % in 2005 to 42% the following year.[28] After storming the domestic market, shanzhai cell phones quickly spilled over the national border. Internationally, 40 million were sold in 2007, and that number jumped to 61 million two years later.[29] By the time of our interview in 2013, Chen quoted from memory that the percentage of the international brands had been hovering at around 50 percent.[30] And in the following year, the combination of shanzhai and Chinese national brands (Lenovo, Huawei, MI, and Coolpad) which mainly used the chipsets by Spreadtrum and its main competitor, Mediatek, is believed to have taken up more than 70% of the market share, leaving the remaining market share to brands such as Apple and Samsung.[31]

"We got the money to develop 3G. But in fact, 80% of the 3G money went to 2.5G. The profit from the 2.5G shanzhai phones sustained us into the 3G era."[32] Chen summarized what they did with a familiar Chinese phrase—"display the sheep's head and sell dog's meat"—basically a bait-and-switch tactic.

This tactic was crucial to Spreadtrum's initial success because, he explains, "The 3G standard was set in 2000, and the market was projected to be ready in 2006. But in fact, it wasn't until 2009 that 3G started to turn a profit, as we had correctly estimated. A handful of companies came up with the 3G technology years before the market was actually ready, and they starved to death." Chen summarized his second lesson as: "The rule of doubling the predicted market development time."[33]

It was long past dinner time in his WestSummit waiting room where we were, but Chen continued to tick off the many advantages of returning to start a business in China, oblivious to the still jammed traffic obscured by the smog nine stories below: "Had we not returned, we would not have known the 'mountain stronghold' market for cell phones, let alone 'co-founding' it." But now, he said, Spreadtrum was expanding beyond the shanzhai market in two opposite directions. On the lower end of the market, it was expanding to other developing countries; going upmarket, it was also moving to name brands.

By Chen's estimate, money goes three to four times further in China than in the US. This is primarily because the Chinese are more willing to bite the bullet if they have to, working longer hours and for lower pay, especially in a crunch. This is critical for a start-up, Chen states, because it allows the leadership to learn and adjust its course of action. "The market is going to decide the fate of the product. Customer demand is the only reason for your survival, and the leader of a company is first of all its primary salesperson. The leader's vision is the ceiling of the company." He pointed out, "This is not a profession for normal people. You're a god if you succeed, and you're a demon if you lose." He emphasized the importance of being aggressive by distinguishing "wolf culture"—aggressively hunting for what you want—from "farmers' culture," being bound to the land and defined by it. In the newly created shanzhai market, he said, he did not wait around for the market to decide if he was a god or a demon. He blanketed the market with the best and cheapest to make the market speak for him. And the resulting increase of Spreadtrum's market share made it a success story.

As he talks, he whisks out a large-screen Samsung phone from his pocket. Lifting up his thick eyeglasses with one hand, he manipulates the phone with the other, shows me the functions onscreen, and says: "Our chips are now adopted by Samsung and other name brands."

The 80 Percent Product and 100 Percent Commitment

Curious about the shanzhai phones on the market, I decide to buy one in a store.

My search turns out to be fruitless. No one admits to selling shanzhai phones; they offer only "authentic" or "smuggled" phones. I tell my friend Luo Qun, a Tsinghua-educated engineer who has worked for over a decade in the US high-tech industry and is now a returnee herself. To help me buy an actual shanzhai phone, Luo gives me a basic education: "You go into a store, tell them your price range, and describe the basic functions you need." The description, she explains, allows them to sell a shanzhai phone: "Shanzhai phones usually have the rock-bottom prices at around 500 yuan (about US $90). Sometimes you can bargain it down to 300 (US $50). But nowadays you'd be surprised how many functions a shanzhai can offer you."

Then she explains why walking into a store and directly asking for a shanzhai phone does not produce results. She informed me that some vendors create unethical shanzhai phones. In her view, the only unconscionable thing about shanzhai phones is when the names and external designs of other brands are imitated, work that is done after the product has left the chip producers like Spreadtrum. "It's like fake Gucci bags to meet the market demand. So you'll find shanzhai phones with a brand like 'Nolcia,' that looks like a Nokia. But, you know, the chip inside is not stolen property—it's a local product." In almost half the cases, this chip is the TD-SCDMA chip designed and made by Spreadtrum. She adds, "I think the proper way to do it is to sell these phones as their own brands in the low-end market as some companies are starting to do."[34]

But if the speed, the technology, the new business model, and the low profit margin successfully brought the shanzhai phones to market, Spreadtrum's failure in the MP3 race once again reminded Chen that the best technology doesn't necessarily guarantee success. Halfway into Spreadtrum's R & D process to come up with a perfect chip for a dual-track 128kps MP3 player, Mediatek flooded the market with far inferior single-track 64kps, a technology Spreadtrum already had. Spreadtrum's leadership had misjudged the market demand. The market didn't need a perfect product. It wanted what Chen termed the 80 percent product: "If the market could tolerate 60 percent quality and be perfectly happy with 80 percent, then we should offer products of 80 percent quality in the shortest possible time." Moreover, the speed of competition combined with the extremely low profit margin makes it almost impossible to produce a high-end product in a timely fashion. Thus, in Chen's view, this 80 percent rule has proven to be a golden rule for the Chinese market.[35]

From across the ocean in Silicon Valley, Yang Zheng closely monitored the intensity of the competition in China's hi-tech world. In his late 30s, Yang is almost a generation younger than Chen. A Chinese college graduate trained in MIT's graduate program, Yang decided not to return to China at the end of his training.[36] China presents many attractions as well as concerns for these Western-trained engineers. A native of Shanghai and a resident of Silicon Valley, Yang follows the success of people such as Chen from the US, and not without envy. "Chen helped take OmniVision public, and this helped him with Spreadtrum's IPO, as it was his second time," he says, indicating that some of the skills and experiences can be transferred advantageously to China. But the brutal pace of coming up with new products and the aggressive business model are some of the most important factors at play in the Chinese market: "The big companies controlling the high-end market are under constant pressure to stay ahead, and innovation in the field moves very fast. The faster the smaller companies can come up with a new technology or a business model, the faster they will catch up to the big guys. Business agility is the most powerful weapon start-ups have, and something lacking in most big companies." While this is also true elsewhere, its intensity is turned up a notch with the competition among different groups of powerful players in a market that hosts the major foreign companies, domestic Chinese companies, and shanzhai phone producers.

The temptation to try his hand in such a market is palpable, but Yang adds that he still struggles with the question of returning to China: "It's never an easy personal choice to relocate back to China." He notes that uprooting a family to live in major cities in China means sacrificing quality of life in many fundamental ways, such as quality of air, water, food and other everyday necessities. It also means jumping into business competition with cutthroat intensity. Such a move is particularly difficult when a family is still young and comfortably established in the States, as is the case with his family.

A "Foreign Exile" in Silicon Valley

If Chen had similar concerns he did not talk about them. China energizes him, empowers him, and gives his work meaning and even historical significance. For him, returning to China was never just an option. It was his professional destiny. Born at about the same time as the PRC, his formative years coincided with the nation's most tumultuous years of revolution. What he jocularly calls his two "gap years" of laboring in the countryside after high school taught him lessons of human endurance and resilience, and the subsequent twelve years at Tsinghua gave him the best training the nation could offer. His time in the US, which he likens to a "foreign exile," is often mentioned in the same breath as his countryside experience. To him, all these were preparations for the show he is producing in his home country and Spreadtrum is his debut. But if America was not his final destination, his decade in Silicon Valley (1990–2000) prepared him well for his China run.

Ming, Chen's former student at Tsinghua and still a close friend, says Silicon Valley is a place "where everyone tries something new every day." While there as a researcher, she saw people succeed and fail, and, having failed, try again. "No one tells you something cannot be done or that what you've dreamed up is wrong. It's like the last place that remains in the Wild West. Everything is possible if you try, and nothing is too formidable."

It was this spirit that snatched Chen out of the doldrums of working as a senior engineer at National Semiconductor, his first job in America after his postdoctoral work at the University of Illinois and Stanford. "I could finish a week's work in a day or two," he says of the job. The rest of the time, he took advantage of the company's incentive of a $2,000 award for every invention, and started a streak that now stands at 41 patents.[37] But one day a former advisor from Tsinghua asked him to recommend someone who could help solve a bipolar IC problem for his company, Opus. Bipolar IC is an integrated circuit chip with a bipolar transmission. This chip is the brain of an electronic device in that it does all the thinking and calculation. "IC design was then Silicon Valley's star profession, and I happened to have just finished reading a book on bipolar IC design." Chen smiles as he recalls the coincidence. With the newly acquired knowledge from the book he decided to recommend himself for the job.

He spent two to three weeks on night shifts at Opus and cracked the problem. This gave him a newfound confidence and fame in Chinese techie circles. Another request for help came from another company: OmniVision. "I had never heard of the CMOS image sensor chip that the new company was to design, but it sounded challenging enough to pique my interest. And this time the new company wanted me to take charge of research and development as a co-founder." Recalling that event, he laughs at his own naïveté: "I had no idea about either the CMOS image sensor or what being a co-founder really meant, but I liked the idea of saving the time it would take to find a new and more challenging job than what I had at National Semiconductor. So I was 'incorporated'."

A CMOS (Complementary Metal Oxide Semiconductor) image sensor converts light into electric charges and processes it into electronic signals. It is often compared to a similar technology called CCD (Charged Coupled Device) for its advantageous low-power consumption, high-speed read-out, and lower fabrication cost. Compared to the more dominant CCD technology of the time, the new generation of CMOS chips transformed the playing field of image sensing by being light, fast, small, and far cheaper to manufacture, although some might challenge the last feature. Its major application at the time was in computers. OmniVision's chip enabled the industry to mass-produce micro-cameras for the computer. Many credited OmniVision with replacing 20 years of Japanese domination in CCD technology with CMOS technology, thus ushering in a new era with cameras in small electronic devices such as computers, cell phones, security surveillance devices, and, later, many medical imaging devices. By the time OmniVision went public on Nasdaq, it had taken 50 percent of the world's market share.

OmniVision had scored an unlikely success. A clear underdog with only a handful of key researchers and dismal funding compared to the big name competitors, OmniVision suddenly found itself on top of the world. That success taught Chen that one did not need to blindly worship a name brand company. "In Silicon Valley," Chen tells the new generation of entrepreneurs at Golden Seed, "95 percent of the innovations were done by small companies and start-ups. And the big companies? They rely on their brand and their established channels and scale of production. They buy new technologies from the smaller companies for blood infusion."

Yang Zheng recalls that around the time he arrived in Silicon Valley in the late 1990s, the digital camera was still in its infancy, and the quality was so bad that no one really used that function. OmniVision was able to focus on developing something no one had ever imagined before and the chip became an indispensable part of devices. It was an incredible feat to not only accomplish that but also to continue that winning streak throughout the decade. The fact that its products have been adopted by Apple and Samsung, the leading brands in the market, is the best testament to the high quality of the product because, Yang notes, "Apple uses only the best." To Yang, that success is due to vision combined with a technology that stands the test of time and competition: "It took guts and skill to create something entirely new, but more importantly, it takes passion and perseverance to carry it through once the new product is established. But the hardest thing is to come up with something no one has ever thought of before." Chen himself says as much, suggesting that coming up with something new is crucial: "Once you show something can be made, others can make the same thing in different ways."

Give Your Dream a Chance

"When you listen to a start-up story, you need to try to understand the mechanism behind it," Chen tells his Golden Seed admirers in another lecture as he advises a new generation of entrepreneurs who have been admitted to the incubator for the promise shown by their new start-ups. "You don't get into the business to be trendy or to make money. There are many easier ways to make money. Creating a start-up is hardly a job for humans. It demands too much to be worth the money you make. So you might ask: When should I try to build a start-up? Well, I'd say that when you have this idea that refuses to go away, an idea that comes back again and again and keeps you up night after night, then you put down your feet and give your dream a chance."

Yang agrees with Chen about the motivation for beginning a start-up. "These guys are not in it for the money," Yang Zheng says of Chen Datong and his cohort, reminding me that Chen had already had his first big pot of gold after OmniVision went public in 2000. Chen, Yang says, learned to dream in America, and with his unwavering perseverance he has realized his dream in his home country now.

"By returning to China and doing what I'm doing, I feel I'm living a life that's fuller than many lifetimes combined for many people," Chen tells me. Like pushing the limits of physical pain, Chen says that he takes delight in conquering challenges and finding ways to expand the limits of his life experiences. Chen feels he is quite fortunate because his personal life seems to have worked out, too. He and his wife lead a lifestyle that many of their friends consider to be simple and frugal to a fault. His only daughter is now enrolled at a major US university. His wife is an activist in disaster relief efforts and has spent months on end, for example, helping out in an AIDS village in China. And when Hurricane Katrina hit, she flew to the US to help out with the relief effort there. The plain and unassuming social image of the couple has no trace of even the dust of the gold that is in their pots.

From the early stage of reform in 1989 when Chen left Chi-

na for his "foreign exile," China has now come to be seen by
Chen as a platform for him to succeed and to expand in relation
to both domestic and foreign competition. Aside from the initial
support he needed to get Spreadtrum off the ground—notwithstanding the wavering support that almost derailed the whole
plan—the government does not seem to have interfered in its
actual operations, perhaps due to a lack of expertise. And when
the company went public in 2007, it was able to operate according to a set of international rules.[38]

To Chen, the significance of his endeavor goes beyond being able to set up and run a successful company with his core
collaborators in their home country. He sees the historical significance of his enterprise in light of Chinese history. And here
is where national and personal ambitions begin to blur. As mentioned above, Chen sets Spreadtrum's story—his own China story—side by side with the story of the CCP. Each validates and
reaffirms the other, and the growing status and competitiveness
of the company thus becomes part of the nation's success as well.
Chen expresses his pride at the end of our conversation by saying, "China now is on track to lead the world market. Huawei [a
major Chinese brand] sells to emerging markets in Africa and
South Asia. Now 20 percent of Spreadtrum's products serve the
high-end market and major Western brands no longer have exclusive rights to call the shots."

While the personal costs of this high speed chase to the top
of the world market deter many would-be returnees like Yang,
they do not shake Chen's resolve in furthering his pursuit. He
wants to make an impact on society by continuing to lead. He
consciously documents the milestones of each lesson learned and
each milestone of progress made. And he tells and retells his story in the context of China's long search for modernity and its
effort to reclaim its past glory and rise as a world power. As part
of his new endeavor he started a new company with a new set
of priorities, a venture capital company called WestSummit. He
envisions supporting a new generation of technocrats who will
lead and transform society. And he puts considerable thought
and effort into pushing toward this vision.

WestSummit finally takes him off the start-up treadmill and

provides a perfect platform and more flexible schedule for pursuing his new vision.[39] The company focuses on funding high-tech companies around the globe with assets at or above $5 million and already with sizable sales. With the combination of his experience in the hi-tech world and available capital, he now has the power to influence a new generation both inside and outside of China by choosing which start-ups to fund. But more importantly, he wants to help cultivate a new generation of civic and technical leaders for China.

"The greatest invention of Silicon Valley is people," he says to me in one interview. "Technology depends on people. Leadership needs the right people. And if you have the right people and bind them together, you can take the team anywhere and succeed in what you do." To Chen, this Silicon Valley take-away is far more significant than any technological invention. Of his own technological breakthroughs and inventions, he explains, "I learned them from books, not from the authorities. Chinese are smart," he says of all those he had worked with, "and they can eat bitterness [meaning they can endure hardship]." With this insight, his newest endeavor is to build a leadership team from among the two groups he sees in China's current hi-tech workforce.

For the homegrown start-up founders and would-be founders still in college in China, he makes time to give lectures every month. He doesn't want them to be dazzled by his technical and financial success in both China and America. Instead, he reminds them of their social responsibility while encouraging them to excel in technology. Distinguishing tangible social contributions such as taxes from intangible contributions such as leadership, he advises the young founders in his Golden Seed lecture series that to make a social contribution you need to first have a viable and even successful enterprise. "The market is there, you have to conquer it with your revolutionary technology."

For the returnees, he emphasizes that their advanced knowledge of technology alone is not the key to success. In the lecture notes on leadership he sent me, Chen lists three major advantages a returnee might have: (1) they have seen the world and thus are not intimidated by the Fortune 500 companies; (2) they are equipped with top-notch technology; and (3) they have usually worked successfully on a product. This third advantage, he says, marks an important difference from academics because these returnees know how hard it is to turn an idea into a product. Chen lists two major disadvantages: (1) they have not sold things; and (2) they usually have not led or managed a team. Chen, who believes that the leader of a company is first of all its number-one salesperson, reminds them that returnee leaders still have much to learn in their home country.[40]

Having seen and succeeded in the world himself, he is trying to bring these two groups of leaders together. His confidence and vision for China are clearly shared by the young entrepreneurs at the Golden Seed lecture I attended. When the lecture ends, I ask Liu Yun in the next seat whether he plans to go to the US someday. He looks at me, pauses, and says: "The question now is why would I go to the US when we have teachers and advisors like Chen Datong?"

As a member of China's first post-Cultural Revolution college class, Chen started his personal pursuit of a dream at the same time as his country re-embarked on its modernization effort in the late 1970s. Today, his personal path to success has become inseparable from that of his country. China gave Chen a decade of training at the nation's top engineering university, Silicon Valley taught him additional skills amid creative people and the idea that dreams can be real, and now China once again provides a stage from which he will compete globally. Looking back, Chen takes a page from the Chinese Communist Party's historical narrative to define his own success. If he first left the country in search of a path as a pioneer, he is now taking his home country along the path of joining the world as a formidable player. "We are not witnessing history," he corrected me by repeating my comment, "we are *making* history." In the context of the CCP's legacy, Chen could easily have used one of Chairman Mao's slogans to describe his story: "Continue the Revolution."

Chapter 5 Notes

1. The first hi-tech start-up incubator was set up in Wuhan, Hebei Province in 1987. By the end of 2012, there were 1,200 across the country "incubating" 70,000 start-ups. Over a 25-year span, 45,000 start-ups have "graduated" from these incubators to become companies. Of this total, 9,000 were started by returnees. Seventy-three percent of those working in these companies are college graduates, and among these, 20,000 are returnees (see "A Summary of Incubators" on the web page of Torch High Technology Industry Development Center, Ministry of Science and Technology, at http://www.chinatorch.gov.cn/fhq/gaishu/201312/b6cbd28149864ff98d382f88491be03b.shtml [December 22, 2013]).

2. In the spring of 2013, I had a few interviews with Chen Datong in his office in the downtown financial center of Beijing. In addition to these interviews, I was able to participate in this session at the incubator.

3. This is explained in the IEEE Xplore Digital Library, "A single-chip image sensor for mobile applications is realized in a standard 0.35um CMOS technology" (http://ieeexplore.ieee.org/xpl/login.jsp?tp=&arnumber=1088114&url=http%3A%2F%2Fieeexplore.ieee.org%2Fiel5%2F4%2F23646%2F01088114.pdf%3Farnumber%3D1088114). OmniVision's own site explains this technology: "Omnivision's first-to-market OmniBSI pixel technology literally turns the imaging world upside down. OmniBAI uses backside illumination light sensing to deliver ultra high-quality sensors among the highest resolutions available and in the smallest form factors. OmniBSI's ability to continue shrinking pixel geometries without compromising image quality or low-light performance makes it ideal for all mobile applications, as well as any other market looking for ultra-compact, high-performance camera solutions" (http://www.ovt.com/technologies/).

4. Chen emailed me PowerPoints for a series of his lectures, including the one for this particular lecture. I am therefore able to quote from it verbatim. He has also sent me a series of articles he wrote for his lecture series at Tsinghua, providing me more context and information.

5. On September 20, 2012, the *Wall Street Journal* published an article titled "The Venture Capital Secret: 3 Out of 4 Start-ups Fail." An online magazine, *ehEurope*, published an article on February 19, 2014, titled "The Chinese Startup Scene, a Gold Rush or Minefield?" It highlights the following: ". . . only 1% of startups in China actually survive" (see http://www.entrepreneur-handbook.co.uk/the-chinese-startup-scene-a-gold-rush-or-minefield/).

6. The Cayman Islands are one of the most attractive tax havens. Barry Naughton notes that, aside from Hong Kong and Taiwan, "Tax havens such as the Cayman and British Virgin islands are also popular choices for incorporation of high-technology start-up businesses in China itself. Creation of an offshore vehicle facilitates the financing of new ventures by both Chinese and offshore investors" (*The Chinese Economy: Transitions and Growth* [MIT Press, Cambridge, 2007], 413).

7. This is Chen's retelling of the birth of the Chinese Communist Party (CCP). Here I summarize the notes he gave me: On July 23, 1921, the first

CCP meeting was held in Shanghai. The meeting was interrupted by the arrival of the police, and the representatives had to hold the last day of the meeting on a boat on the South Lake in Jiaxing, Zhenjiang Province. The meeting produced the first party statement and resolution, and elected the central bureau. Chen is here comparing the meeting of the Spreadtrum founders to this last day of the CCP meeting on the boat after they were evacuated from Shanghai. But more importantly, he used the revolutionary nature and the ultimate success of the CCP in the subsequent decades to foreshadow the future of Spreadtrum.

8. Barry Naughton points out, "It was not until 1999 that Chinese firms were given across-the-board support to enter high-technology fields as private firms and start-ups. In place of the earlier policy of only favoring large SOEs (State-Own-Enterprises), the government now supports virtually all technologically advanced enterprises, including small, private start-ups and technology-intensive spin-offs from schools and research institutes. In an important shift, instead of seeing private firms as rivals with publicly owned enterprises, these firms are now viewed as 'national' enterprises . . . The nature for the support for high tech firms has changed as well. Increasingly, the government provides a kind of across-the-board support for domestic enterprises designated 'high technology.' Tax breaks, access to low-interest credit lines, preference in procurement decisions, and other kinds of regulatory preference or relief are all used . . ." *The Chinese Economy: Transitions and Growth* (MIT Press, 2007:361).

9. According to *The Statistical Abstract of China* (SAC), in 2004, among the total of 1.16 million individuals engaged in R&D (or a full-time equivalent), 920,000 were scientists or engineers. In the section "Human Capital Resource Base," Barry Naughton calls for caution in using these statistics because the qualities of the graduates from different colleges could differ greatly. However, he says of the returnees: "At the top of the skills pyramid an important role is played by Chinese who have studied abroad . . . over the years training of Chinese scientists and engineers in advanced degrees overseas has contributed an enormous amount to the growth of China's human resource base" (p. 361). Returnees have played a disproportionately large role in fostering new high-tech start-ups and upgrading educational institutions. Even when students do not return, they play a role in connecting domestic scientists and engineers to international networks of research and innovation (Wang Huiyao, *Returnee Time* 海归时代, Central Translation Press, 2005).

10. This set of numbers comes from Chen's memory, and is slightly different from the data listed in the first note in the Introduction of the current book. But since the discrepancy does not undercut the idea that it was a highly selective process in 1977, I decided to use Chen's memory here.

11. "N" refers to an unknown quantity, and many in the tech world like to use it to indicate "many."

12. Barry Naughton (2007) points out that, from 2001 on, China had more inclusive technology-promotion policies, and that they began to have success in attracting joint-venture IC factories to the mainland (see p. 369).

13. A detailed discussion of Spreadtrum's investors can be found in Claudio Petti, ed., *Technological Entrepreneurship in China: How Does It Work?* (Edward Elgar Publishing Limited, 2012), 132.

14. WCDMA stands for Wideband Code Division Multiple Access. This is a third-generation (3G) wireless standard which utilizes one 5 MHz channel for both voice and data, initially offering data speeds up to 384 Kbps. WCDMA was the 3G technology used in the US by AT&T and T-Mobile.

15. The second round of funding was from a combination of sources: Fortune, Pacific Venture Partners, Vertex, HuaHong, etc. It arrived in 2003.

16. The third round of funding came from more than 30 investors. The leading investors included: New Enterprise Association (NEA), Fortunetech Investment Fund, Pacific Venture Group, Vertex, Legend Capital, and Hua-Hong International.

17. ABB stands for Analog Baseband chip; DBB stands for Digital Baseband chip; and PWIC stands for Polarization and Wavelength Independent Coupler. A comprehensive list of different rounds of investors can be found in the chapter on Spreadtrum in Claudio Petti, ed., *Technological Entrepreneurship in China: How Does it Work?* (Edward Elgar Publishing Limited, 2012).

18. TD-SCDMA stands for Time Division Synchronous Code Division Multiple Access. GSM stands for Global System for Mobile Communication.

19. This is the kind of technology innovation that Clayton Christensen termed "disruptive innovation." In his book *The Innovator's Dilemma* (Harper Business Essentials, 2003), he explains the evolution of the size of computer disk drives and points out: "The most important disruptive technologies were the architectural innovations that shrunk the size of the drives—from 14-inch diameter disks to diameters of 8, 5.25, and 3.5 inches and then from 2.5 to 1.8 inches" (p. 16). But more than the size, these innovations are usually smaller, cheaper, and offer higher performance.

20. GSM stands for Global System for Mobile Communications.

21. The award was Best Market Performance, China Chip Award, given by the Ministry of Information Industry. The name China Chip is a play on homophones of *zhongguoxin*, where the same reading can be written in different ways to mean different things: 中国心 "China heart" or 中国芯 "China Chip."

22. Although the origins for the term "Mountain Stronghold" remain fuzzy, it is clear that it refers to a rebel that disrupts normal market operations. Clayton Christensen has coined the term "disruptive innovation" for this kind of practice: "Generally disruptive innovations were technologically straightforward, consisting of off-the-shelf components put together in a product architecture that was often simpler than prior approaches" (*Innovator's Dilemma*, Harper Business Essentials, 2003), 16.

23. This is generally regarded as a watershed year, or the year the shanzhai cell phones made an impact on the market. Mediatek is generally credited to be the first to flood the Chinese market with low-cost turnkey cell phones in 2004. Therefore, Chen Datong claims to be the "co-founder" of the shanzhai market when the market really opened up the following year.

24. By 2006, one source of information shows a pie chart indicating that Mediatek supplied 40 percent of the chips sold in China, and Spreadtrum supplied 10 percent (see http://www.pcpop.com/doc/0/287/287519_all.shtml).

25. In *The Innovator's Dilemma*, Clayton Christensen distinguishes "sustaining technology" and "disruptive technology." The former is often made up of incremental improvements by a leading company, and the latter is usually a much more radical change (see his note 23). "These [referring to the disruptive technologies] were the changes that toppled the industry's leaders" (p. 15).

But here, once again, instead of defining his strategic success in the context of Western business management, Chen likens it to the CCP's strategy that eventually enabled them to take over China from the domination of GMD (Guomintang, the Nationalists). CCP was founded in 1921, almost a decade later than GMD. In its struggle to take over control of the country from the GMD, the CCP aligned themselves with the peasants and workers (the largest section of the population), establishing their bases in remote areas far from the GMD power centers in the major cities. In the final year of the civil war (1945–1949), the Communist PLA (People's Liberation Army) swept through the country from these bases in the remote areas, taking over major cities and establishing the People's Republic of China in 1949. Associating with this grassroots aspect of CCP history conveys a positive image, and it evokes a sense of indigenousness, elements Chen clearly wants Spreadtrum to be associated with.

26. Zhaoqin Xiang 向兆琹, "Research Found Sales of Chinese Shanzhai Cell Phones Reached 145 Million Sets 查示09年中国山寨手机售量达 1.45 亿部," *Tencent*, http://tech.qq.com/a/20091106/000228.htm (November 6, 2009 [accessed November 2015]).

27. Chinese national companies started to hit the market in 1998, taking less than 3 percent of the market share in 1999. Yet their number has doubled every year since then: 7.5 percent in 2000, 15 percent in 2001, 30 percent in 2002, and by 2003, the Chinese national brands took half of the overall market share, according to an article co-authored by Jiutang Pan 潘九堂, Liu Hui 刘, and Yuan Quan 袁泉: "An Analysis of the Origins and the Competitiveness of Shenzhen's Shanzhai Cell phone Production System 深圳山寨手机生体系的起源和争力分析" at http://www.ide.go.jp/English/Publish/Download/Jrp/pdf/156_ch1.pdf (accessed November 2015). The sudden drop of the domestic brand's market share was the topic of discussion in many articles across the media. A search on the 2005 Chinese domestic cell phone market share will yield many online pieces.

28. An online article by Xijing Cao 曹希敬 and Hu Weijia 胡佳, "The Evolution and Indication of China's Shanzhai Cell Phones 中国山寨手机的演及启示" has this data: To meet the market demand, the production of "turn-key" cell phones increased from 37 million in 2005 to more than 100 million in 2006, taking 40 percent of the overall sales in China that year. From 2006 to 2008, the *shanzhai* market surged again, its production increasing from 150 million in 2007 to nearly 200 million in 2008, about one third of the total

cell phone production that year, pushing down the market shares of established brands. Nokia, for example, reduced its average price in China from 102 Euro per handset in 2006 to 62 Euro in 2009 (see http://doc.mbalib.com/view/df9ee3c7dcf974aa05d746f23b4894bb.html [accessed November 2015]).

29. Shanzhai handset shipment in domestic and export markets, millions of units:

	2005	2006	2007	2008	2009	2010	2011	2012	2013	2014
Domst.	31	41	51	40	33	24	22	18	17	16
Export	6	14	40	61	112	204	233	195	175	152
Total	37	55	90	101	145	228	255	213	192	167

Source: Jiutang Pan 潘九堂, Liu Hui 刘, and Yuan Quan 袁泉, "An Analysis of the Origins and the Competitiveness of Shenzhen's Shanzhai Cell phone Production System 深圳山寨手机生体系的起源和争力分析" at http://www.ide.go.jp/English/Publish/Download/Jrp/pdf/156_ch1.pdf (accessed November 2015).

30. A later source in 2014 indicates that domestically produced smart phones have taken up 70 percent of the domestic market (see "2014 Chinese-made Smart Phone Production Increases by 25%, Intensifying the Clash Between Domestic and Foreign Brands 2014 中国智能机出量或增 25％，激化土洋对攻," *Enterprise General Situation* 企业概况, Panda (http://www.panda.cn/SJTCMS/html/PandaJT/pandajt201310/83820548.asp [accessed March 29, 2016]).

31. An online source, "Research Institute of Chinese Ministry of Industrial Information: Domestic Cell Phone Sales of National Brands Dropped 25% from Last Year 工信部研究院：国手机量同比下降25.4％," noted that in October 2014, national brand cell phones took up 78.2 percent of overall sales (see http://tech.sina.com.cn/t/2014-11-13/doc-icesifvw7344203.shtml [accessed November of 2015]).

32. Clayton Christensen (2003) has demonstrated that the forecast the sustaining technologies does not work for disruptive technologies in the section of *The Innovator's Dilemma* entitled, "Forecasting Markets or Sustaining Versus Disruptive Technologies."

33. The difficulty of discovering new markets is well-explained by Clayton Christensen (2003): "Markets that do not exist cannot be analyzed . . . not only are the market applications for disruptive technologies unknown at the time of their development, they are unknowable . . . few have any theoretical or practical training in how to discover markets that do not yet exist . . . The processes demand crisply quantified information when none exists, accurate estimates of financial returns when neither revenues nor costs can be known, and management according to detailed plans and budgets that cannot be formulated." Thus, Christensen concludes: "Given the powerful first-mover advantages at stake, however, managers confronting disruptive technologies need to get out of their laboratories and focus groups and directly create knowledge about new customers and new applications through discovery-driven expeditions in the market place" (*Innovator's Dilemma*, chapter 7).

34. The emergence of the shanzhai phones provoked a debate both at

home and abroad over the ethics of such a business practice. In 2009, when the threat of shanzhai phones was felt outside of China, with its third-year rapid increase to 61 million sets sold abroad, articles in *Forbes* and *The New York Times* started to pick up on the debate. The 2009 *Forbes* article, "China's Black Market Boom" by Gady Epstein, reads: "By flooding the market with their own 'brands' that are not exact copies, with labels and advertising that are also intentionally slight alterations, the bandit production lines are muddling a fundamental question of intellectual property: Where do counterfeits end and genuine products begin?" *Forbes* (http://www.forbes.com/global/2009/0216/014.html [February 16, 2009]).

Within China, "shanzhai" became a catch phrase applied to many things other than cell phones. One master's thesis in law, available online (Zhang Lei, Yangzhou University, Civil Law, 2011), said of the word: "It can be said to have entered every aspect of our lives. People have different opinions about shanzhai—is it a kind of grass roots knowledge, an innovation, or plagiarism? People of different opinions are having a hard time to convince each other" (http://211.68.184.6/KCMS/detail/detailall.aspx?filename=1013128857.nh&dbcode=CMFD&dbname=CMFD2013 [site no longer available on March 29, 2016]). A Chinese site on legal issues (http://wenku.baidu.com/view/5422d175f46527d3240ce0ca.html?re=view) had a long article devoted to legal analysis of the shanzhai phones. The abstract of this article reads in part: "The shanzhai cell phones did not violate the patented inventions because they circumvented that, but they violated design patents and Unfair Competition Act because their shape and trademark resembled those of legal brands'" (retrieval sometimes difficult). Another site summarizes the debate on both sides: Lei Xie, ed. "Understanding Shanzhai: Is It Grassroots Spirit or a Killer of Innovation? 解"山寨"：草根精神抑或新"于"？" *News of the Communist Party of China*, http://theory.people.com.cn/GB/40555/8610266.html (December 31, 2008).

35. For Chen's 80 percent rule, Clayton Christensen puts it this way (*Innovator's Dilemma*, 2003): The products "offered less of what customers in established markets wanted and so could rarely be initially employed there. They offer a different package of attributes valued only in emerging markets remote from, and unimportant to, the mainstream" (p. 16). In this case, the mainstream would be the established foreign IT companies.

36. Yang Zheng is a personal friend. I have had many email exchanges with him over the course of writing this chapter during 2013. As a member of the successful elite in the hi-tech culture (he himself has co-founded and sold start-ups), he is also very knowledgeable about the culture, and insightful in analyzing the different choices these elites make.

37. The search in the USPTO Patent Full-Text and Image Database yielded 41 patents either by Chen Datong alone or with Chen as co-inventor, as of March 2015 (see http://patft.uspto.gov/netacgi/nph-Parser?Sect1=PTO2&Sect2=HITOFF&p=1&u=%2Fnetahtml%2FPTO%2Fsearch-bool.html&r=0&f=S&l=50&TERM1=chen+datong&FIELD1=&co1=AND&TERM2=&FIELD2=&d=PTXT).

38. Spreadtrum went public on June 27, 2007, with the IPO code of SPR, raising 125.9 million, and went for public trading on NASDAQ the following day.

39. "The company is always 18 months away from bankruptcy." This is what Chen Datong learned from his Spreadtrum experience.

40. These principles are taken from the files Chen sent me.

An Age of Citizenship

Studies of China's more than three-decade-old policy of "Reform and Opening" often emphasize the wealth and power that have been generated as a result. Some of these studies stress the positive sides of reform and others the negative. Most, however, tend to emphasize the importance, for better or worse, of the State, the party, and top-down authority as the motive force behind all change. Much less emphasized, and much less understood, is the role ordinary citizens have played in fostering sweeping societal change. It is precisely that role—the role of the citizen—that has been the underlying theme throughout this book. The point is obviously not that China today is somehow a thoroughly free and open society. Nor is the point that individual rights are either protected or wholly respected in contemporary Chinese society. Rather, the point is that with accelerating momentum throughout the Reform Era, regardless of what actors within the State apparatus might have intended, ordinary people are increasingly taking control of their own lives, shaping their country's destiny, and assuming responsibilities most of us in any modern society associate with citizenship.

This notion of citizenship, manifested in the sorts of returnee endeavors described throughout this book, may seem pedestrian to the average American or Western European reader. People in advanced industrial democracies have for generations conducted business through markets, participated in the creation of children's libraries for reading and play, engaged in general aviation, taken up environmental causes, or even fought for evolving definitions of gender and marriage rights. For many observers in the West, these are just normal aspects of modern social life. In that sense, none of the activities described in this book are extraordinary.

What is extraordinary, however, is that they are taking place today in China. That, more than anything else, is the point. In contemporary China, along with the rising GDP has come a social discourse increasingly resembling, again for better or for worse, the one found in many modern societies. Chinese people today are living their lives as modern citizens, taking up fights over issues about which they are passionate, trying as best as they can to improve their own lives, and, in fact, often striving to realize a greater good for all of society. To that end, often in very subtle ways, these citizens, loyal though they may be, challenge the status quo, challenge the government, and challenge the bureaucracy. They are neither radicals, revolutionaries, nor dissidents, but they are consistently pushing the boundaries of what is tolerable politically and socially in order to strive for something better. In that sense, though they may see themselves as apolitical, they are potent agents for change. That they have been able to do this—again, regardless of what actors at the top of China's governmental hierarchy may or may not have intended—signifies the real achievement of Chinese reform.

As they assert this newfound awareness of their powers of agency, citizens routinely bump up against a recalcitrant system. The sociologist Li Yinhe has long argued that the Chinese legal system and state bureaucracy are always at least a few years behind the newest trends in society, a lag that often causes friction. And, as party leadership is accustomed to flexing its muscles when challenged, it tends toward draconian measures to maintain control. Everybody in China today recognizes and celebrates that rapid change is underway. But nobody is entirely satisfied. In the context of change, the political leadership strives to maintain control, while the citizenry pushes for greater autonomy. Many of the returnees described in this book find themselves negotiating between those two impulses: the desire to maintain societal order, and the equally strong desire to push for a very different tomorrow. Step by step, and ever so delicately, they negotiate a path toward progress through everyday acts.

Becoming Returnees

I have chosen to focus on returnees not because they are more important than any other social group. Rather, it is because they have the opportunity to live out a dream that, at least in the early stages of reform, was shared across wide swaths of Chinese society. Back in the 1980s, when citizens across the country were desperate to realize a better life, these returnees were the lucky few who had a chance to get out and leave for what everybody assumed were far better prospects abroad. Improving one's life within China then was hard and slow going. The prospects seemed dim, and there were no established rules to facilitate more entrepreneurial pathways. It is true that even under such conditions (and foreshadowing what returnees would do decades later), farmers still managed somehow to start rural firms and some city dwellers opened small private businesses, but it was very tough going. In the absence of clear ownership and regulation, trying something new meant venturing into unknown territory and constantly looking over your shoulder for fear of becoming the target of an imminent crackdown. Thus, during this first period of reform, the most desirable entrepreneurial act of all involved leaving the country. The few who could access such opportunities jumped at them. Many of those who left never returned. But, as the years went on, some people did venture back home. Most of the returnees described in this collection returned to China before it was fashionable to do so and before prospects for success were particularly promising. Yet they returned nonetheless, and did so at once inspired by their American education and experience and determined to realize a new vision for their own society.

In returning, they were—and still are, of course—trying to make money, trying to get ahead, and trying to gain in social status. But they also sought to make a difference in society. They sought to leave a lasting mark. This is where self-interest and patriotism converge. Just like other citizens, these entrepreneurs don't necessarily agree with the government or support all aspects of the government's social system, but they understand now that they have the power to shape their social environment. They can be agents of change professionally and perhaps even politically. Fundamentally, they no longer have to be subjects. This book shows how these acts of citizenship and change have been enacted on a daily basis.

Like comparable transformations in other countries in recent decades, China's transformation has been neither clean nor pure. It has not always been fair, nor has it been without victims. This book does not gloss over that gritty reality, and indeed it cannot. As much as it embodies China's transformation, the experience of these returnees also reflects the country's problems and failings. In the transformed social landscape, the returnees described here can now be considered part of China's new elite and their astonishing achievements part and parcel of the disparity and inequality of the contemporary Chinese milieu. The returnees' success in some ways hinges on connections and privileged access to power. Thus, along with their success in exercising agency and pushing for social change, the returnees themselves have become part of the problem. And this too is a way the returnees have become part of the country's contemporary narrative of change.

As new members of China's social establishment, the returnees participate in and rely upon social networks that are often close-knit and overlapping. Many of these networks can be traced back to old school ties. Indeed, as one respondent would introduce me to another, often by way of shared social and university connections, my own access was facilitated through these networks. Thus, while the industries these returnees work in might seem quite disparate, the individuals themselves are often closely connected, sharing common acquaintances, common early life experiences, common educational achievements abroad, and common perspectives now that they have returned to China. That too is a theme of this book. For all the differences in their styles and preferences, and for all the richness and diversity of their professional experiences, these individuals embody more than just randomly selected data points. They are connected socially, experientially, and, in some sense, aspirationally. That is, they are striving for a common set of societal goals.

Making the Unconventional Conventional

In the years that have passed since my last round of interviews in 2013, the returnees documented in this book have continued to lead in their fields while making news, branching out, taking new routes, and trying out new ideas.

When, at the end of 2014, sociologist Li Yinhe publicly acknowledged that her long-time life partner since the death of her husband is a transgendered man, many observers in China were shocked. Suddenly, the scholar and the subject of her scholarship converged. Previously, Li Yinhe's studies of "unconventional" sexuality and her advocacy of same-sex marriage rights had been considered avant-garde, but also, in a sense, abstract. Now her own lifestyle appeared shockingly new, real, and concrete.

After the news became public, Li uncharacteristically granted a series of major interviews, first to the domestic Chinese media and later to foreign outlets, including *The New York Times*. The *Times* article appeared on March 6, 2015, months after Li's story was first made public in Chinese media, but at a significant time in the US.[1] Here the same-sex marriage debate was in its final stages and the Supreme Court ruled in its favor just a little over two months later on June 26. Li explained that she had not timed her revelation to coincide with heightened passions in the US debate, but the impending Supreme Court decision had generated a spike in the long-simmering rumor in China that Li was a closeted lesbian. So she decided to bring it out in the open.

In media appearances at the time, Li was in the company of her partner, Daxia, who tends to be the more talkative of the two. In their unvarnished and down to earth manner, they presented their story of basic human attraction and love. The photo in *The New York Times* perfectly captured this tone. The couple sat on the edge of a bed by a large bright window with their heads turned backward slightly toward the camera. If it were not for Daxia's spiky short hair, nothing would appear even remotely out of the ordinary. The room is indistinguishable from the thousands of others one would expect to find in the apartment of a happily married, middle-aged couple. And the couple themselves, with rounded faces softened by age, seem relaxed, at peace, and happy in the home they have made together.

Responses to Li Yinhe's "coming out" varied and were often emotionally charged. Li's detractors seized upon the news as evidence of her lack of credibility as a scholar. Her devoted supporters hailed her for her bravery. Many others simply did not know how to react. A Chinese academic noted to me, "She has done some really good things, but now she is simply weird. I don't know what to make of her personal life. But as long as she continues to do what she's always done, I suppose her personal life shouldn't be that big a deal." In a way that is perhaps reflective of this comment, many people seem to find Li Yinhe's story somewhat strange, but they are not interested in telling somebody else how to live his or her life. Nor, at least in this case, is the government telling people what to think. In the end, Li is practicing what she advocates: the idea that one is entitled to a lifestyle of one's own choice, and that people are entitled to have their own opinions. And, in a manner that would have been unthinkable twenty or thirty years ago, this public story of an unusual lifestyle just passed like a ripple through a sea of social change.

It was in such circumstances Li came to speak at Brown University in April 2015. Once again, politics intervened. On her first day in America she learned that government security forces had detained five Chinese feminist activists. Their supposed crime was planning to disseminate anti–sexual harassment information on public transportation systems. Outraged, Li Yinhe spent her first morning in the US writing an essay and posting it immediately on her blog site. In the essay, she affirmed China's relatively positive status globally with respect to gender equality. According to data she cited, the nation ranks 28th globally on gender equality measures out of 100 nations surveyed and China, at the time Li was writing, was drafting a new law against domestic violence. Li then openly wondered how this very same government could choose to appear as if it was siding with abusers by arresting those fighting for gender equality. The implication was clear: the government had inadvertently wandered down a path that served neither its own interests nor its reputational status. She concluded by offering a way out. The government should just release the women, tacitly admit it erred, and move on. As Li noted, it is always better to immediately admit and correct an error than to deny and conceal. Instead of directly attacking the government's foolishness in detaining these activists, Li chose to emphasize her country's positive achievements and hold the government to the high standard it had already achieved for itself. Of course her comments contained more than a bit of irony, and they were written in a tongue-in-cheek manner. But still, for Li, the ultimate goal is not to shame or score points, but to keep moving toward the goal of achieving social progress. If the government would be willing to concede after being faced with reasonable and loyal remonstration, then Li would be willing to extend an olive branch as well. She would gladly pledge her loyalty as long as the march toward social progress continued unimpeded.

Though this particular blog entry is relatively mild, Li was on tenterhooks (a new phrase she loves to use) the whole day after it was posted. She worried that either her blog would be deleted by the government, or worse, that the entire site might be shut down. But as the day wore on, thousands of followers began visiting her site with numbers mounting hour by hour. By the next morning, Sina, one of the most powerful Chinese media companies, had decided to recommend the blog on its own site. With this recommendation, Li's blog went viral, reaching tens of thousands of readers, many of whom left online comments.

The Internet has become a powerful facilitator of Li Yinhe's public engagement with state authorities. It enables her followers to exercise their rights as citizens at least to a degree, for they can choose for themselves what to read, what comments to make in response, and what policies to support and reject. Needless to say, all these behaviors still have limits in China. But no one seems to know where the red lines are drawn, not even the authorities. The public forum of the Internet has itself become a game of push and pull.

As Li explained to me during those days, "Sina was also waiting to see the government's reaction to my blog." The online media company did not want to court sanctions by recommending her posting too soon. In the end, Li felt that she once again squeaked past a threshold of government sanction and the government's firewall was breached just a bit more. This is typical of Li Yinhe in action: she is determined not to be stopped, all the more so while in the United States.

Li Yinhe's talks at Brown attracted large audiences, with some people driving hours to see her. At one talk, she told the audience that the pressure on the Chinese LGBT community comes mainly from society and families and not so much from religious groups. As LGBT group members are primarily young and well-educated, they actively use social media to promote social understanding of their lifestyle. Conflicts sometimes flare up and remain, but society in general is becoming more accepting of this social group. For these reasons, Li told her audience, she remains optimistic.

Li's upbeat attitude is not wholly unfounded. Lifestyle choices and sexual preferences, including Li's "unconventional" ones as measured in terms of traditional Chinese practices, are increasingly becoming a matter of individual prerogative in contemporary China. Undoubtedly, many of the individuals making these choices still face varying degrees of social and familial pressure amidst an atmosphere of controversy and conflict. Yet these same individuals are now able in most cases to openly strive, if they so choose, for more widespread social acceptance. Coinciding with Li's trip, news arrived on her last day in the States that the five activists had been released. "That's the right thing to do," she commented on the government's decision.

Becoming Capitalists

Engaging government authorities is also a necessity for entrepreneurs in China's increasingly globalized economy. When I asked the IT magnate Chen Datong for an update since our last interview in Beijing, Chen's first response was that the government was finally putting money in the right place in support of the hi-tech industry. "A ten percent annual growth rate is not sustainable over the long run," he writes in his email, referring to the year-on-year GDP growth numbers the government had been posting. He commented that the recent economic slowdown was a sign of the economy maturing and consolidating. He bases his confidence in the future on two of his latest business endeavors:

> 1) After its acquisition by the Tsinghua Group in September 2013, Spreadtrum (the first company that Chen Datong helped found when he returned to China) has made major progress.[2] Its employees have increased to 4,000 from 1,600 in just two years, its market share continues to increase, and its major clients now include Samsung and HTC.
>
> 2) Chen's Hua Capital Management, an investment firm that targets the semiconductor sector, has made two major acquisitions, the Silicon Valley-based companies OmniVision and ISSI. OmniVision is actually a firm that Chen helped found 20 years ago and sold. Now, Hua Capital has repurchased it for $1.9 billion. The second firm, ISSI, was acquired for $840 million. Chen is now planning to take both of these firms public in China in the coming years.

With the purchase of OmniVision, Chen has come full circle. America is no longer the "land of foreign exile" it was for him a quarter century ago. Now he has repatriated to China the fruits of his long sojourn in the US.

Chen views his personal achievement in a larger context. For him, it is a harbinger of things to come not just for him but for his country. Chen cites as a watershed the Chinese government's June 2014 report, "Outline for Furthering the Development of the National IC Industry."[3] After this document was released, Chen commented in the online *Chinese Electronic Newspaper* that the highlight of this document is the government's proposal to establish a national fund (Sina IC fund) to foster the IC (integrated circuit) industry.[4] As he later explains to me, the most significant aspect of the document is the government's commitment to finance tech ventures in a new manner. Instead of following its traditional approach of taking a 100 percent stake in ventures and thus crowding out private investors, the State will instead take only 25 percent, leaving the rest to the private sector to make up. The goal here is to insure that funded ventures respond to market pressures rather than political ones. To Chen, this could mark the start of a virtuous cycle by which companies will have to answer to tech-savvy, profit-focused private investors instead of political bosses who may have money but lack both market sophistication and meaningful pressure to realize profit and manage risk. As a result, funded businesses within their own management teams will have to hire skilled experts instead of cronies or political appointees. And if these businesses are able to survive market competition, they will emerge with a certain amount of supportive government regulation and funding. In the year and a half after this initiative was implemented, Chen points out that there have been more mergers and acquisitions between Chinese and American companies than the combined number during the five previous years.

As Chen sees it, this emerging trend of Chinese acquisitions of US high-tech companies was preceded by major shifts in the market. "All signs are indicating that the semiconductor industry has gradually morphed from infancy to maturity. Shanzhai phones and their demand for commodity integrated circuits are now a thing of the past. Telecommunications and the Internet are industries that started in the West, but now their market is shifting to China, and many of the earliest corporate leaders are being edged out by homegrown Chinese stars," Chen notes. He then points to three reasons for such a shift. "First, China has the right market type for this stage of the industry's development; second, China has a disproportionately large number of consumers for hi-tech telecommunications products; and third, as chip design has a built-in security code, domestic companies have advantages when it comes to national security issues. As a result, many foreign companies have retreated in just the past two years." As our conversation continues, he ticks off company names: TI, ADI, Siemens, and Freescale, all of which have stopped making mobile phone chips. Broadcom and Marvell stopped making phone chips just last year. The only one left is Qualcomm.

Smartphone and tablet makers in China have also gone through a major shift in the past two short years. "Former giants such as Motorola, Siemens, and Nokia have stopped making smartphones for the Chinese market," Chen observes. When Chen speaks, company names fly around like cards in an intricate game, shuffled and reshuffled as the narrative shifts. "Once Chinese makers started to produce similar products with lower profit margins, these giants retreated or withdrew. Among the major smartphone suppliers in the Chinese market today, aside from Apple and Samsung, all the others are home-grown Chinese companies, the top three of which are Xiaomi, OPPO, and Huawei." Then Chen zeroes in on the main targets in the game. "Apple is still strong and hard to beat, occupying more than ten percent of total global market share. With its well-established ecosystem and a large and loyal fan base, it is going to remain strong for a while. For the Android system, the top company in China is still Samsung. But Samsung's position in China will likely be squeezed by domestic brands, as it already shows signs of retreat. These domestic Chinese companies are not only driving the competition domestically, they are also turning up the heat outside China's borders in India and elsewhere. This situation was completely unimaginable just two years before." Two years ago it would have been equally unimaginable for Chen that Chinese investment companies would be buying Western tech firms. Today, Hua Capital and WestSummit—Chen's two investment companies—have made that a reality and are getting almost weekly inquiries from the West for more acquisition deals. The whole game has shifted.

One such acquisition that Chen helped broker involved the takeover of Segway by a new Chinese company called Nine-bot. WestSummit and Xiaomi co-invested in the acquisition in March 2015.[5] By owning Segway, Chen says, Ninebot became the number one player worldwide in the two-wheeled, self-bal-ancing electric transporation vehicle market. Ninebot also gained an instant brand name.[6]

While most in the field would agree that Ninebot bene-fits from the Segway acquisition, not everyone agrees on exactly why. Ninebot's cofounder, Wang Ye, sees in the deal the patents that come with Segway, intellectual property that strengthens Ninebot's products.[7] A *Time* magazine article published during the acquisition adds a different piece of the picture. "Just seven months ago," according to the article, "Segway filed a trade com-plaint accusing Ninebot and other Chinese companies of violat-ing its patents. Their products indeed resemble one another, but Ninebot has insisted it 'independently owns its intellectual prop-erty.'"[8] Given such circumstances, it appears that the acquisition of the Segway brand name resolves the intellectual property dis-pute, removes the cloud of patent infringement, and validates Ninebot's products. For Chen Datong, the purchase was a logical result of a price gap. "Notwithstanding the fact that Segway had been struggling almost from its inception, when a Segway is sold for $7,000–$8,000 and a similar Ninebot product is 7,000–8,000 RMB (about $1,300–$1,500), it is hard for the Western compa-ny to compete. Furthermore, Ninebot also came out with a sim-plified version of the product that's sold for a fraction of that at 1,999 RMB, appealing to a far larger section of the population."

Chen extols the domestic market's dynamism, supportive government regulation, and a smarter government approach to public-private financing as some of the key reasons for the tre-mendous progress made in the hi-tech industry in the past two years. As a counter-example, he points to misguided policies of the past that led to some of the most serious problems today, in-cluding economic slowdown and corruption. These wrong-headed policies include the government's indiscriminate funding of real estate early in the Reform Era coupled with the use of GDP as the chief criterion to measure officials' achievements. When a government employee can use the government's money to immediately increase GDP in his area, Chen explains, the official does not have to consider the longer-term market return for the investment. As a result, useless projects have been built for the sake of bumping up the short-term GDP number, personal pockets have been lined with government money, and officials have simply gotten a transfer when things did not work out.

Using the solar photovoltaics (PV) industry as an example, Chen makes his case. When the solar PV industry was found to be profitable, local government officials swarmed in to build solar PV manufacturing facilities. Without overall planning or a system in place to streamline production and use of solar PV products, huge waste resulted, and in many cases the environment suffered. Worse yet, when technical problems occurred, local officials did not have the expertise to fix them. But as many political satirists point out today, the irony is that those who worked to boost the GDP have been promoted and those who worked on the environment have been passed by. In Chen's view, had the government started from the beginning with what it is doing now—investing heavily in the hi-tech industry and doing so with built-in incentives to compete on market terms—it would not have to run a sweeping anti-corruption campaign.

Today, Chen is heartened by the government's interest in and new funding structure for the hi-tech market. Looking ahead, Chen sees China's competitive edge in transitioning to a hi-tech workforce, and he thinks China is ready for such a transition. A migrant worker's average monthly pay has increased from 500–600 RMB a decade ago to 3,000–4,000 RMB today, three times higher than that of Vietnam's. In 2000, the fabless design house portion of the semiconductor industry had a total revenue of less than $100 million. Now it is nearly $20 billion (more than 200 times growth in 15 years). "Just two years ago," Chen said, "it was impossible to buy a foreign company even for $100–200 million. Now we are gobbling up US companies worth $840 million and $1.9 billion. The accumulated wealth of the country has made these large transactions much easier now." Chen foresees a wave of IPOs (initial public offerings) by Silicon Valley companies now trying to list on domestic Chinese stock markets, and by extension, raise China-based funds. Hua Capital's acquisitions of OmniVision and ISSI are just part of this trend.[9]

This trend of acquiring American companies seems poised to spread to the general aviation industry. Despite the economic slowdown and anti-corruption campaign in China, Wu Zhendong's companies have fared quite well.[10] Responding to my request for an update, he lists airplane sales worth millions of dollars.[11] Wu acknowledges that because of "anti-corruption, a soft economy, the stock market crash and the fluctuation of oil prices," 2016 will likely be a slow year. But he is philosophical. "All things move in cycles," he says. Looking beyond the immediate horizon, he foresees expansion into a number of new areas.

One such step would be to "acquire fifty percent of Seacore shares this year or next." Seacor Holdings, Inc., a publicly listed US company, was Avion Pacific's partner in founding Asian Sky Group (ASG) in March 2012. Back in 2012 when ASG's general manager at the time, Jeffrey Lowe, was asked what he and his company do, he responded that he was "merely an educator, i.e., trying to educate the market as to the benefits of purchasing a business jet." But, he continued, "[T]he Chinese get it now."[12] Now, ASG's website lists a range of consulting services, including "Aircraft Sales, Completion Management, Operation Oversight, Luxury Charter, Special Projects, and transactional advisory," among others.[13] "I have high hopes for this company and great plans for it to grow into one of the finest professional aviation advisers for big corporations, general aviation operators, jet owners, and aircraft OEMs (original equipment manufacturers)," Wu writes in an email.

Wu's high hopes extend to his other company, King's Aviation, which he says is growing "slowly but surely."[14] King's Aviation now has three helicopters working on power line inspection and he plans to increase the variety of their services. Plans are also in the works for expansion, including the addition of fixed-wing aircraft, and the establishment of a possible MRO (maintenance, repair, and overhaul) joint venture to support after-sales services.

For the two entrepreneurs, Chen Datong and Wu Zhendong, their measure of success is the achievement of true market leadership in highly competitive global environments. Both have shown skill in leveraging Chinese market power to achieve competitive advantage in the global arena. Early in the reform era, the defining characteristics of the Chinese market were low wages and a large customer base. Chen Datong leveraged that ten years ago by successfully unleashing the shanzhai phone wave. In Wu's case, when he started his general aviation business in the early 1990s, his competitive advantage basically involved selling a CAAC connection to foreign companies trying to gain a foothold in the Chinese market. The foreign company gave the money and Wu gave them CAAC.

Today China's population is larger and, more important, vastly richer. Correspondingly, both Chen Datong and Wu Zhendong are now playing an entirely different commercial game. Chen Datong has moved into the realm of investment and finance by buying foreign companies and helping to launch their IPOs in China. Zhendong is fully in his element now that the general aviation sector is finally hitting its stride in the Chinese economy. "Aviation is an exciting business because it is high profile," he wrote at the end of 2015. "The big money and high risk keep the adrenaline going, and the hard work dealing with customers in the hope of attaining long-term return ropes one in." What Zhendong said two years before still holds true: With China behind him, people will have to deal with him. This will only be more true as this China becomes even richer.

As heads of their own businesses, both Chen Datong and Wu Zhendong have control over their firms' operations and planning, just as business leaders do in developed countries. This kind of private entrepreneurial control was unthinkable under pre-reform Maoist rule, or even in the early Reform Era for that matter. Politicians back then issued top-down directives to industrial sectors where firm-level managers worked to fulfill production quotas. It is a perfect irony that entrepreneurs like Chen and Wu who grew up in a Communist China that antagonized capitalists have now become distinguished capitalists themselves. Moreover, the very success of their businesses gives them the credentials and gravitas to claim a privileged voice in contemporary China, a voice that has increasing influence over how the country is run.

Lifting the Dome: Combatting Air Pollution

Other citizens, even those who lack the capital and prowess of wealthy entrepreneurs, are also making their voices heard. Environmental activists are an important example. At the end of February 2015, *Under the Dome*, a documentary by a well-known China Central Television newscaster, Chai Jing, caused a sensation. After becoming available for free online, it had millions of views on the first day. Moreover, it initially garnered public praise from the then-newly-appointed Chinese environmental minister, Chen Jining, himself a returnee from England. In the film, Chai started her narrative by describing how her own young daughter lived as a virtual prisoner at home because of the toxic air quality outside. Chai went on to lay out the details of China's severe pollution problem. She pointed the finger at polluters large and small; explained the various health risks of such serious pollution; and urged citizens to take action to hold the government and polluters accountable. The tone of the film was more a wakeup call than a cry for revolution. Indeed, it offers hope by comparing China's current pollution to the London smog of 1952, implying that even the worst smog problems can be alleviated with determination and focused action, much as Britain had done. A week after *Under the Dome* was released online, however, Internet links to the film were abruptly blocked. The environmental minister too backed off his support for the film, instead emphasizing that no other nation had spent as lavishly as China on environmental protection.[15]

Months after domestic Chinese access to Chai's documentary was blocked, and right around the time of the Paris climate conference, *The New York Times* ran a series of articles on the alarming levels of air pollution in China. An article in early December 2015 reported Beijing's first issuance of a citywide red alert for bad air quality. According to the emergency air-pollution response system announced by the government in 2013, a red alert goes into effect if there is a prediction that the air quality index PM2.5 (a measure of fine particulate matter) will stay above 200 for more than 72 hours. In the US as well, PM2.5 levels of 200 or above are considered "very unhealthy," and a level of 301–500 is "hazardous."[16] When the Chinese government issued a red alert for Beijing in December 2015, the PM2.5 reading in the city that day stood at 253 by 7 p.m. Red alerts of the kind issued in Beijing that day require school and office closings, as well as severe restrictions on automobile use.

The picture is bleak, no doubt. In many ways none of these problems is entirely new, as they have been well documented and widely reported for years. For anyone who has lived for an extended period in China in the past few years, particularly in northern Chinese cities like Beijing, horrifying PM2.5 levels have been frequent if not routine occurrences. When my family and I lived in Beijing in 2012–2013, my two teenage sons experienced cancellations of after-school activities due to air quality issues on several occasions. One evening we had a dinner appointment in the city with a friend who worked at the US embassy in Beijing. The friend called hours before our appointment to cancel for fear of leaving his air-purifier-equipped room. He said the embassy's reading of the PM 2.5 index had passed 800, and could be as high as 1,000. But he had no way to ascertain the actual number, as there were no instruments that would reliably measure a number that high. At that time, anyone with a VPN (Virtual Private Network) could go online and get two simultaneous, albeit often different, readings of the air pollution index, one from the US embassy and one from the Chinese government's official site. The official Chinese reading, which is made public, is usually much lower than the US embassy measurement. At that time in 2012–2013, there was no official red alert system in place to sound the alarm.

So if the problem is not new and perhaps not substantially worse than it was back in 2012–2013, why now are we hearing more about it through various media *and* the Chinese government's official red alert system?

Incrementally over time, reports and studies within China have educated the public about the health risks from contaminated food sources, algal blooms in waterways, soil degradation near industrial bases, and beach pollution along China's coastline. Chai Jing's documentary has pieced together many of these problems into a cohesive and damning narrative of a dire environmental situation nationwide. The combination of rising awareness and persistent campaigns by activists has catalyzed a society-wide sense of urgency to face the problem.

Lately, the Chinese government's legislature has seemed willing to respond, at least in some ways. One result was the aforementioned emergency air-pollution response system announced in 2013 and implemented for the first time in 2015. Whether it was by choice or by fiat, the Chinese government has now made verbal commitments to play its part in confronting environmental problems. In September 2015, Chinese President Xi Jinping and US President Obama signed the US-China Joint Presidential Statement on Climate Change. The Statement paved the road to the Paris climate agreement concluded three months later. Through this combination of heightened awareness of the need for environmental protection and increased willingness on the part of the government to respond, the sounds of alarm have come through much louder than ever before.

The fight against environmental degradation takes many forms, and involves many kinds of activists. Liao Xiaoyi, China's most decorated veteran environmentalist, is still working on educating people to befriend nature and reduce their consumption of the earth's dwindling resources. These days, while occasionally still promoting her recycling and energy saving projects, Liao devotes most of her energy to restructuring and transforming her Global Village of Beijing into a social enterprise.

The direction Liao has taken over the past two years seems to have distanced her from the global environmental movement. In the euphoria at the conclusion of the Paris climate deal in late 2015, Liao and her organization were conspicuously absent, a strange outcome given their long history of activism on the Chinese environmental scene. In a series of BBC Chinese reports on the Paris conference, one article recounted the active role played by Chinese NGOs at the conference, but made no reference to Liao or her NGO, the Global Village of Beijing.[17] Liao seems to have lost faith in crunching numbers on emissions levels and temperature rise. In talks and interviews prior to the Paris conference, she did not mention this highly anticipated event. Instead, she promoted her theory of "lehe," or what she terms "life in harmony." When asked why she retreated to the background after having made an impact on China's environmental protection efforts, she responds that she is now leading a new trend by emphasizing the work of rebuilding a sustainable society in the countryside.[18] It is in the countryside and within rural communities, she asserts, that the soul of the nation resides. This, she thinks, is the ultimate answer for environmental protection and rejuvenation.[19]

It may take years for Liao's approach to show results, and there is no precedent for achieving success along the path upon which she has embarked. But such is the nature of this reform from its inception. When Chinese people set out to realize the Four Modernizations over thirty years ago, they had no clear roadmap and no well-defined goal. What they did want was something better than had been experienced during the decades of Maoist revolution. Is the China they see today, then, what they set out to build over three decades ago? There is wealth and there is optimism, but along with both have come less positive things most Chinese never signed up for. Environmental pollution is one. In 1949, Chairman Mao resolutely declared that the Chinese people have stood up. Today, Chinese people await the day when they can declare, "We have fully arrived."

Defining "Progress"

In the course of national development, at least in the case of China, it turns out that envisioning progress and arriving at a better place are more complicated than declaring independence. How exactly to measure a country's progress—and according to which standards—has long been a topic of debate among experts and observers of many stripes. As mentioned earlier, China had for a long time focused on raising its GDP. But the practice of using GDP as the main measure of national strength and well-being has been questioned ever since the term was conceived by Simon Kuznets in the 1930s. Indeed, Kuznets himself warned that "the welfare of a nation can scarcely be inferred from a measure of national income." Decades later, Robert F. Kennedy described GDP as something that "measures everything . . . except that which makes life worthwhile." And, today, when IT magnate Chen Datong criticizes the Chinese government's overemphasis on GDP growth, he mirrors Nobel Prize-winning economist Joseph Stiglitz's statement, "What you measure affects what you do. If you have the wrong metrics, you strive for the wrong things."

If GDP alone is not enough to measure national progress, what other metrics should be considered? Which other targets and measures would contribute to the realization of a better future? Whether for nation or citizen, what does a "better future" or "better life" exactly mean? Having achieved success along the conventional standards of education, wealth, and status, the returnees described in this book find themselves turning to these deeper questions of value, meaning, and what in the end really counts, whether for themselves or their society.

Sonya, who has been managing Avion's Shanghai office for years, decided in 2015 to relocate back to her old haunts in Los Angeles. A single mother of a teenage daughter, she wants her child to attend secondary school in the US, and she wants to take some time off to reassess her own career goals and life priorities. While managing Avion Shanghai, Sonya—a slight woman just over five feet tall—taught helicopter search-and-rescue techniques to law enforcement agencies and developed extensive ties to a vast network of emergency responders. Now in the US on hiatus from Avion, she has used these ties to become involved with China-based volunteer rescue teams, civil society groups that have formed to provide disaster relief and special rescue services at a moment's notice whether in China or abroad. These teams, as Sonya explains, are composed mostly of successful Chinese professionals who also happen to be outdoor adventurers. From Los Angeles, she remotely manages their missions to many different parts of China, as well as to less-developed parts of Asia. During the May 2015 Nepal earthquake, for example, Sonya was on the US West Coast, yet she was able to manage fundraising efforts and coordinate rescue operations for Chinese volunteer responders. Sonya views the dedication and effort of these volunteers as some of the most uplifting examples of human altruism she has ever witnessed. She hopes one day to join them on some of their missions once her daughter enters college.

Reassessing career and life priorities has been an increasingly present theme for Peeka's founders as well. By the end of 2015, only Yi and her family were still living in China full time, though they too spent summers in the US to permit the children to attend camp and educational programs. Hu Birong is living in Silicon Valley with her two children. Aidong and her children moved to Birong's neighborhood in the fall of 2013, and then Ming and her children followed in 2014. Though dedicated to fostering continued change in China, all three women recognize that that change process is moving too slowly to meet their children's needs. They mean this in the most fundamental sense. The US still offers a far healthier environment than China. America offers cleaner air, safer food, and, less concretely, a society rooted in more established norms and values. The women all feel that children in their most formative years need to be surrounded by values such as these. In a sense, they reflect the views of many Chinese, who, though deeply proud of their country's achievements, worry that during these decades of rapid growth, Chinese society has lost its moral compass.

Yi, Birong, and Ming are still as engaged in Peeka as ever. Through Yi and chief librarian Cheng Xin (their "boots on the ground"), all three founders, wherever they happen to reside, operate and collaborate in the same way they always have, but now through extensive online communication. To them, the Peeka libraries are still the point-of-entry for participation in their country's campaign for a better future. For them, a better future means a whole generation of children growing up as thinkers, inventors, and well-educated global citizens rather than test-takers. That's what these women will continue to work on. But what they do goes beyond just the immediate—and decidedly critical—task of providing education for young children. They see their work in China as representative of something bigger: civic duty. For them, this is about responsible citizenship. As modern citizens of China, they are committed to keeping this status and playing this role in their native country.

This latest cohort of returnees have set themselves apart from their predecessors in history. Earlier returnees in the nineteenth and first half of the twentieth centuries were treated by Chinese society as one-off cases of extraordinariness. This individual might have been considered an extraordinary hero, and that one an extraordinary traitor. But they were each *sui generis* and removed from normal society. The current generation of returnees, however, is different. It has now become a coherent group, a social class embedded within broader Chinese society, and deeply enmeshed in the Chinese social establishment. As an elite class, so to speak, current returnees wield considerable power to influence and shape their country's future. At the same time, they are in a position to accumulate substantial personal fortunes. Their work, for better or worse, is now fully integrated into the country's tableau of change. The returnees have become key players within the establishment, catalysts for some of China's most important social transformations, and, in many cases, key interlocutors in their country's interactions with the broader global community.

Some of these returnees initially came home because they anticipated one day hearing their children ask, "Where were you during China's most rapid growth in the modern era?" Now, that day has arrived, and many of their children do in fact ask that question. Many of these returnees are proud to say that they have been in the thick of it all. In Chen Datong's words, they have helped make history. But the children now may also follow up with another equally valid and perhaps more poignant question: "Where were you when I was growing up?" It is this kind of question that reminds returnees of the personal sacrifices they made on their road to success. In the end, though, these individuals can take comfort in the fact that they have delivered to their children a world with more opportunities, more comfort, and more wealth. They have delivered a world in which their children and the children of many others will be able to exercise powers of citizenship that just decades earlier would have been unimaginable. Many, many problems still remain to be solved. But for this entirely new generation of Chinese—the children of the returnees—citizenship has become their birthright.

Conclusion Notes

1. See Andrew Jacobs, "Sex Expert's Secret's Out, and China is Open to It," *The New York Times* (March 6, 2015).

2. The acquisition took place in September 2013.

3. The Outline is available on the web page of the Ministry of Industry and Information Technology of the People's Republic of China Website 中华人民共和国工业和信息化部 网页, at http://www.miit.gov.cn/n11293472/n11293832/n11294042/n12876231/16048026.html.

4. Chen's article, "Government Foundation Allied with Private Capital, Holding Up the Sky for the IC Industry. 政府基金与民间本联手 中国IC工业擎起一片天," *Chinese Electronics News* 中国电子报 at http://semi.cena.com.cn/2014-07/04/content_231514.htm (July 4, 2014 [accessed March 29, 2016]).

5. Ninebot is a startup founded in 2012 by a couple of graduates from Beihang University (Beijing University of Aeronautics and Astronautics), China's preeminent aeronautic and aviation institution.

6. Segway's purchase by a Chinese buyer is widely reported. *USA Today*, for example, carried an article on April 15, 2015: Kelvin Chan, "Segway Bought by Chinese Company," *USA Today* (April 15, 2015).

7. In addition to gaining a brand name, the co-founder of the company Wang Ye explained in one exclusive interview that the main advantage of acquiring Segway is the many patent rights that have come with the purchase. Heitaiyi, "Cofounder of Ninebot Wang Ye: Buying Segway is Only a Start. Future Products Will Include Domestic Robots That Can Walk and Work. Ninebot 联合始人王野：收只是始，未要做能走路能干活的家用机器人" http://36kr.com/p/531894.html (April 16, 2014 [accessed February 2016]).

8. Jack Linshi, "Why This Chinese Startup Just Bought a Company Americans Love to Ridicule." *Time* online magazine, Business section, at http://time.com/3822962/segway-ninebot-china/ (April 15, 2015 [accessed February 2016]).

9. At the same time these conversations with Chen Datong took place, *The New York Times* carried an article about Alibaba's purchase of a Hong Kong newspaper, the *South China Morning Post*. Aibaba's executive vice chairman, Joseph C. Tsai, is quoted as saying: "Our business is so rooted in China, and touches so many aspects of the Chinese economy, that when people don't really understand China and have the wrong perception of China, they also have a lot of misconceptions about Alibaba." He then concluded: "What's good for China is also good for Alibaba" (*New York Times*, December 11, 2015). This is very much the sentiment that Chen Datong has expressed time after time in talking about his own success, as well as the growth of the semi-conductor and venture-capital industries (David Barboza, "Alibaba Buying South China Morning Post, Aiming to Influence Media," *The New York Times*, December 11, 2015).

10. Information and quotes directly come from my personal email correspondence with Zhendong in December 2015.

11. They include: "7 Sikorsky S92 helicopters and 8 S76D model SAR (Search and Rescue) helicopters to MOT (Ministry of Transport of China), as well as some business jets and quite a few King Air (propeller aircraft) for special missions." Private email correspondence on December 5, 2015.

12. Mark Nayler, "Ready for Take-off," *Spear's Asia*, Issue 3, 2012/13 at http://www.asianskygroup.com/attachment/news/25/SPEARS-asia-ASG.pdf (February 2016).

13. See the website at http://www.asianskygroup.com/services/aircraft-sales.

14. Here's the original email from Zhendong: "Currently we are conducting power line patrol and inspections using MD Explorer. Kings has 3 helicopters now (still very small) but I am confident that it will grow into more areas (HEMS, Special Training, Aerial Photography, Aerial Mapping, etc). [The] next stage is to introduce fixed wing aircraft into Kings Aviation. We are also in talks with one or two large Western helicopter maintenance companies (they're call MRO—Maintenance Repair and Overhaul) for a possible MRO joint venture in China to support the after-sales service. The rotary wing growth in China is approximately 25% per year, but lacking support services."

15. "Environmental Protection Minister: Has Seen 'Under the Dome', Texted to Thank Chai Jing 环保部部长：己看过《穹之下》，短信感了柴静" *Sohu* online news at http://news.sohu.com/20150301/n409242625.shtml (March 1, 2015 [accessed February 2016]). In another news report, VOA reported "Chinese Environmental Protection Minister Avoids Chai Jing's Smog Film 中国环保部长者会回避柴静霾片," *VOA Chinese* online edition at http://www.voachinese.com/content/china-ban-under-the-dome-20150309/2672842.html (December 14, 2015 [accessed March 29, 2016]).

16. Edward Wong, "Beijing Issues Red Alert Over Air Pollution for the First Time," *New York Times* (December 7, 2015).

17. Xing Li, "Active NGOs at the Paris Climate Conference 巴黎气候

大会上活跃的非政府织," *BBC Chinese* website at http://www.bbc.com/zhongwen/simp/china/2015/12/151208_paris_climate_7_ngo (December 8, 2015 [accessed March 29, 2016]).

18. One such example is "Liao Xiaoyi 'Lehe' Livestyle Gains Resonance 廖晓义'和'生活方式得到同," *Sina* at http://finance.sina.com.cn/hy/20150526/120122268925.shtml (May 26, 2015 [accessed February of 2016]).

19. "Liao Xiaoyi: Big Companies Migrate to the Countryside—Do Not Turn Farmers into Laborers 廖晓义：大公司农村，不要把农民变成打工仔," *Tencent*, no. 14 at http://news.qq.com/zt2010/talkliao/. Another of Liao's talks on the similar topic of turning to the countryside can be found in Huitan Xu 徐会坛, "Liao Xiaoyi: From a Western-styled Environmentalist to an Oriental-style Environmentalist 从西方环保者到东方环保者," *ICixun* at http://www.icixun.com/2015/0305/4855.html, March 5, 2015 [accessed March 29, 2016]).

Bibliography

"On January 29, 1963, Zhou Enlai Proposed the Four Modernizations 1963年1月29日周恩提出四个代化." *News of the Communist Party of China* (on the Four Modernizations). Accessed February 2016. http://cpc.people.com.cn/GB/64162/64165/76621/76651/5289691.html.

"'Opinions on the Admission to the Higher Educations in 1977' As Approved and Issued by the State Department of PRC.国务院批的《于1977年高等学校招生工作的意见." *Chinese Archival Information Web* 中国案网. October 16, 2014. Accessed February 2016. http://www.zgdazxw.com.cn/dagb/2014-10/16/content_70181.htm.

"2010 Hurun Wealth Report: China's Millionaires Numbered 875,000 2010 胡富报告：中国千万富豪达87.5万人." *Sina.* 2010. Accessed October 2015. http://finance.sina.com.cn/leadership/crz/20100401/15347678221.shtml.

"2014 Chinese-Made Smart Phone Production Increases by 25%, Intensifying the Clash Between Domestic and Foreign Brands 2014 中国智能机出量或增25%，激化土洋对攻." *Enterprise General Situation* 企业概况. Panda. Accessed March 29, 2016. http://www.panda.cn/SJTCMS/html/PandaJT/pandajt201310/83820548.asp.

"A Must-Know Acting Career: The Young National Star Actor Chi Zhiqiang 于全国优秀青年男演志强，你不得不知的演艺经历." www.72177.com. April 9, 2015. Accessed February 22, 2016. http://news.72177.com/a/201509/042070640.shtml.

"A Summary of Incubators 孵化器总体概况." *Torch High Technology Industry Development Center, Ministry of Science and Technology.* December 22, 2013. Accessed February 2016. http://www.chinatorch.gov.cn/fhq/gaishu/201312/b6cbd28149864ff98d382f88491be03b.shtml.

"Amended version of the Public Security Administration Punishments Law of the People's Republic of China 中华人民共和国治安管理处法." Accessed October 2015. http://www.gov.cn/flfg/2012-10/26/content_2253934.htm. Also available at Peking University's English website on Chinese laws: http://www.lawinfochina.com/display.aspx?lib=law&id=4549&CGid.

"Announcement of Personal Income Ranking in 31 Provinces: Shanghai First and Beijing Second 个省份人均收入排行公布：上海最高

北京第二." *Xinhua Net.* February 27, 2015. Accessed March 28, 2016. http://news.xinhuanet.com/fortune/2015-02/27/c_1114459674.htm.

"Annual Meeting 2008: Global Citizen Awards." *Clinton Global Initiative.* Accessed March 29, 2016. http://www.clintonglobalinitiative. org/ourmeetings/2008/meeting_annual_GCAwards. asp?Section=OurMeetings&PageTitle=Global%20Citizen%20 Awards%202008

"Chinese Sexologist Sparks Debate on Prostitution." *Radio Netherlands Worldwide.* Accessed October 2015. http://www.rnw.org/archive/ chinese-sexologist-sparks-debate-prostitution.

"Chinese Shanzhai Handset Market Research Report." Accessed November 2015. http://www.qbpc.org.cn/inc/uploads/download/ Mimosa/Shanzhai%20Handset%20Market%20Research-EN%20 20130926.pdf.

"Chopper Market Set to Take Off." *China Daily (Europe)*, April 17, 2013. Accessed October 2015. http://europe.chinadaily.com.cn/ business/2013-04/17/content_16414510.htm.

"Criminal Law of the People's Republic of China 中华人民共和 国刑法." December 22, 2010. Accessed February 22, 2016. www. chinalawedu.com.

"Diagnosis Standards of Psychiatric Diseases中国精神疾病断 准." Accessed October 2015. http://www.wendangwang.com/ doc/46ff7fb9392d3c625b997fc3/2.

"Filed Pursuant to Rule 424(b)(4) Registration No. 333-31926." Accessed February 2016. http://www.nasdaq.com/markets/ipos/filing. ashx?filingid=1222832.

"Gay and Lesbian Rights." *Gallup.* Accessed March 28, 2016. http:// www.gallup.com/poll/1651/gay-lesbian-rights.aspx.

"Heroes of the Environment 2009." *Time.* Accessed March 29, 2016. http://content.time.com/time/specials/packages/0,28757,1924149,00. html.

"Investigation on Ten Yuan Shop Sexual Workers: Women Farmers View Receiving Guests as Working in the Fields 十元店行工作者 查：农视接客种田." *Wangyi News* 网易新. Accessed March 2015. http://news.163.com/12/0427/18/80490KUP00011229_all.html.

"Li Yinhe Promotes Polyamorous Relationships: Avant-Garde View Causes Mass Anger 社会学家李银河憧憬多边恋，一夜情正名

被批." *Tencent.* July 22, 2006. Accessed March 28, 2016. http://news. qq.com/a/20060722/000730.htm.

"Li Yinhe: Decriminalize Prostitution 李银河：淫嫖娼非 罪化." *IFeng.com* "Expert Forum 家." February 11, 2014. Accessed October 2015. http://news.ifeng.com/exclusive/scholar/ detail_2014_02/11/33683553_0.shtml.

"Li Yinhe's Angry Words on Polyandry: I'm Being Demonized 李银河言"多边恋": 我正在被妖化." *People* 人民网. August 16, 2006. Web. March 28, 2016. http://culture.people.com.cn/ GB/22219/4707058.html.

"Liao Xiaoyi 'Lehe' Lifestyle Gains Resonance 廖晓义'和'生活方式 得到同 " *Sina.* May 26, 2015. Accessed February 2016. http://finance. sina.com.cn/hy/20150526/120122268925.shtml.

"Liao Xiaoyi: Big Companies Migrate to the Countryside – Do Not Turn Farmers into Laborers 廖晓义：大公司农村，不要把农民 变成打工仔." *Tencent,* No. 14. Accessed February 2016. http://news. qq.com/zt2010/talkliao/.

"Major Events in the 30 years of China's Reform and Opening: 1978 中国改革放30年大事." *Central Government Information Repository, PRC*中央改革信息. November 6, 2012. Accessed 2016. http://www. reformdata.org/content/20121106/1438.html.

"Major Events in the Development of Beijing Incubators 北 京孵化器发展大事." *Gaoxin Net* 高新网 October 24, 2014. Accessed February 2016. http://www.chinahightech.com/ html/684/2014/1020/936363336353.html.

"Omnivision Rises." *CNN Money.* July 14, 2000. Accessed February 2016. http://money.cnn.com/2000/07/14/deals/omnivision/.

"Peekabook House is No Doubt a Path Blazer in Building Private Libraries." *Workers' Daily.* December 16, 2013. Accessed March 27, 2016. http://acftu.people.com.cn/n/2013/1216/c67502-23845698.html.

"People to know." *Asiaweek*, Vol. 25, No. 38, September 24, 1999.

"Private Libraries: Fraught with Frustrations in Growth 私人，成长 多恼." *Workers' Daily* 工人日报. December 16, 2013. Accessed March 29, 2016. http://acftu.people.com.cn/n/2013/1216/c67502-23845698. html.

"Public Security Administration Punishment Regulations of the People's Republic of China 中华人民共和国治安管理处条例." Accessed October 2015. http://baike.baidu.com/view/33749.htm.

"Public Security Administration Punishments Law of the People's Republic of China 中华人民共和国治安管理处法." Accessed October 2015. http://baike.baidu.com/view/2313510.htm.

"Pushing Open Low-altitude Airspace Across the Nation Next Year, Expect to Have a Huge Big 'Cake' Dropping from Sky." *People's Daily* overseas edition. November 30, 2012.

"Research Institute of Chinese Ministry of Industrial Information: Domestic Cell Phone Sales of National Brands Dropped 25% from Last Year 工信部研究院：国手机量同比下降25.4％." *Sina.* November 13, 2014. Accessed November 2015. http://tech.sina.com. cn/t/2014-11-13/doc-icesifvw7344203.shtml.

"Shanzhai Phones 山寨手机." *Information Archives*文料. http://m.03964.com/read/ca5042f2734b28277d8be7dd.html.

"Why Did Airline Disasters Happen One After Another? Explaining the Nepotism That Tripped the CAAC 空何接踵而至？解倒中国民航的裙系." *The Southern Weekend* 南方周末. May 16, 2002.

"You Thinking of 'Black Flight'? The Fines Can Reach 200,000 Yuan." *General Aviation Net* 通航业网 March 12, 2015. Accessed October 2015. http://www.ethcy.com/htmls/info/thbk/thbkzs/15513.html.

Alitto, Guy. 1986. *The Last Confucian: Liang Shu-Ming and the Chinese Dilemma of Modernity,* 2nd edition. Center for Chinese Studies, UC Berkeley, Book 20. University of California Press.

Asia Pacific Migration Research Network (APMRN), "Migration Issues in the Asia Pacific: Issues Paper from Hong Kong." United Nations Educational, Scientific, and Cultural Organization Website. Accessed October 2015. http://www.unesco.org/most/apmrnwp7.htm.

Asian Sky Group Website. http://www.asianskygroup.com/. Accessed March 29, 2016.

Babbie, Earl. 1987. *The Practice of Social Research (Methodology in Sociology Studies* 社会研究方法). Translated and edited by Li Yinhe. Sichuan People's Publishing House.

Barboza, David. "Alibaba Buying South China Morning Post, Aiming to Influence Media." *The New York Times.* December 11, 2015.

Callahan, William A. 2013. *China Dreams: 20 Visions of the Future.* Oxford University Press.

Cao, Xijing 曹希敬 and Weijia, Hu 胡佳. "The Evolution and Indication of China's Shanzhai Cell Phones 中国山寨手机的演及启示." *Science Technology and Industry* 科技和业. March 2014. MBAlib. Accessed March 29, 2016. http://doc.mbalib.com/view/df9ee3c7dcf974aa05d746f23b4894bb.html.

Carson, Rachel. 1962. *Silent Spring.* Boston: Houghton Mifflin.

Central Intelligence Agency. "The World Factbook." *CIA Library.* Accessed March 28, 2016. https://www.cia.gov/library/publications/the-world-factbook/fields/2018.html.

Cetron, Marvin. "China's Economic Growth Opens Skies for Bizjets." *Professional Pilot.* Accessed March 25, 2015. http://www.propilotmag.com/archives/2011/Oct%2011/A2_China_p1.html.

Chan, Kelvin. "Segway Bought by Chinese Company." *USA Today.* April 15, 2015.

Chen, Datong 大同. "Government Foundation Allied with Private Capital, Holding Up the Sky for the IC Industry 政府基金与民间本联手 中国 IC 工业擎起一片天." *Chinese Electronics News* 中国电子报. July 4, 2014. Accessed March 29, 2016. http://semi.cena.com.cn/2014-07/04/content_231514.htm.

Chen, Datong's patents. *Google Advanced Patent Search.* Accessed February 2016. https://www.google.com/search?tbo=p&tbm=pts&hl=en&q=ininventor:datong+ininventor:chen&num=10#q=ininventor:-datong+ininventor:chen&hl=en&tbm=pts&start=50.

Chi, Susheng 凤生. "Sexual Workers Should be Decriminalized 性工作者非罪化." *360doc.* March 13, 2012. Accessed October 2015. http://www.360doc.com/content/12/0313/19/6795100_194086055.shtml.

Chi, Zhiqiang. "Chi Zhiqiang: If I Were Born 20 Years Later, I Certainly Would Not Have Had to Go to Jail 志强：如果晚生二十年，我一定不会坐牢." *Sina.* January 2, 2009. Accessed March 28, 2016. http://ent.sina.com.cn/m/c/2009-01-02/10432325355.shtml.

Chinese Communist Party Central Committee. "Decision on the Reform of Educational System 中共中央于教育体制改革的决定." The Ministry of Education. May 27, 1985. Accessed February 2016 .http://www.moe.edu.cn/publicfiles/business/htmlfiles/moe/moe_177/200407/2482.html.

Chinese National Health and Family Planning Commission 中华人民共和国国家生和划生育委会. "National Gender Ratio at Birth Shows a 'Six-year Consecutive Decline' 我国出生人口性别比出'六降'." Accessed March 28, 2016. http://www.nhfpc.gov.cn/jtfzs/s3578/201502/ab0ea18da9c34d7789b5957464da51c3.shtml.

Christensen, Clayton. 2003. *The Innovator's Dilemma*. Harper Business Essentials.

Clayton Christensen Institute website (on disruptive technology). Accessed February 2016. http://e360.yale.edu/mobile/feature.msp?id=2782.

Criminal Law of the People's Republic of China. Accessed March 28, 2016. http://www.opbw.org/nat_imp/leg_reg/China/CRIMINAL_LAW.pdf.

Danlan "Pale Blue (or "Blued")," Accessed 2016. http://www.danlan.org/.

Dong, Yuming 董玉明. 2007. *In Step with Reform: Theories of Commercial Law and the Study of Time Management* 与改革同行：经济法理与时间研究. Intellectual Rights Press 知出版社.

Dworkin, Andrea and MacKinnon, Katherine. 1988. *Pornography and Civil Rights: A New Day for Women's Equality*. Accessed October 2015. http://www.nostatusquo.com/ACLU/dworkin/other/ordinance/newday/TOC.htm.

Epstein, Gady. "China's Black Market Boom." *Forbes*. February 6, 2009. Accessed February 2016. http://www.forbes.com/global/2009/0216/014.html.

Fallows, James. 2012. *China Airborne: The Test of China's Future*. Vintage.

Friedman, Thomas. 2005. *The World is Flat*. New York: Farrar, Straus and Giroux.

Gaarder, Jostein. 1996. *Sophie's World: A Novel About the History of Philosophy* (P. Moller, Trans.). New York: Berkley Books.

Gage, Deborah. "The Venture Capital Secret: 3 Out of 4 Start-ups Fail." *The Wall Street Journal*, September 20, 2012.

Gao, Yuanyang. *Gao Yuanyang (Personal) Blog*. Accessed February 2016. http://blog.sina.com.cn/s/blog_593827e90100x4p7.html.

Guo, Jinhui. "Regulations Loosened for Three Types of NGOs, Kick-

ing Off Management System Reform 三类NGO注册条件放，管理
体制改革起步." *Sina.* July 13, 2011. Accessed February 2016. http://
green.sina.com.cn/2011-07-13/135322806932.shtml.

He, Lijin 何利谨 and Chao, Dong 董超. "Returnee Libraries Start to
Increase 海归多起了." *People's Daily Overseas Edition.* May 31, 2013.

Heitaiyi 黑太一. "Cofounder of Ninebot Wang Ye: Buying Segway
is Only a Start. Future Products Will Include Domestic Robots That
Can Walk and Work. Ninebot 联合始人王野：收Segway只是始, 未
要做能走路能干活的家用机器人." *36Kr.* April 16, 2014. Accessed
February 2016. http://36kr.com/p/531894.html.

Huang, Fiona. "The Chinese Startup Scene, a Gold Rush or
Minefield?" *ehEurope,* February 19, 2014. Accessed October 2015.
http://www.entrepreneurhandbook.co.uk/the-chinese-startup-
scene-a-gold-rush-or-minefield/.

Hurun reports. 2013. Accessed October 2015. http://img.hurun.net/
hmec/2013-08-14/201308141028423283.pdf. 2013.

Jacobs, Andrew. "Sex Expert's Secret is Out, and China's Open to
It." *The New York Times.* March 16, 2015. Accessed February 2016.
http://www.nytimes.com/2015/03/07/world/asia/chinese-advocate-
of-sexuality-opens-door-into-her-own-private-life.html?_r=0.

Keck, Elizabeth. 2000. "A WTO Model: Setting Aviation Standards
in China." *China Business Review.* Vol. 27.

Keck, Elizabeth. 2001. "Commercial Aviation Takes Off." *China Business Review,* Vol. 28.

Kelly, Emma. "Pilot Training." *Asian Aviation.* December 10, 2012.
Accessed October 2015. http://www.asianaviation.com/articles/364/
Pilot-Training.

Li, Xing 立行. "Active NGOs at the Paris Climate Conference 巴
黎气候大会上活跃的非政府织." *BBC Chinese Website* 中文网.
December 8, 2015. Accessed February 2016. http://www.bbc.com/
zhongwen/simp/china/2015/12/151208_paris_climate_7_ngo.

Li, Yinhe 李银河 and Hongxia, Zheng 宏霞. 2001. *Grandfather and
Grandson – Case Studies of Chinese Families* 一之 – 中国家庭系个案
研究. Shanghai Cultural Press 上海文化出版社.

Li, Yinhe 李银河 and Xiaobo, Wang. 2005. *Thinkers' Talk – A Collection of Wang Xiaobo and Li Yinhe* 思想者 —— 王小波李银河双人集,
Cultural and Arts Press 文化艺术出版社.

Li, Yinhe 李银河, ed. 1999. *The Debate on the Revision of Marriage Law* 婚姻法修改争. Guangming Daily Press 光明日报出版社.

Li, Yinhe 李银河, ed. 1996; 2007. *Women – The Longest Revolution: Selected Readings on Contemporary Women's Rights Literature* 女 – 最漫长的革命, 当代西方女主义理精选. Joint Press 三联店. 1996; Chinese Women's Press 中国女出版社, 2007.

Li, Yinhe 李银河, ed. 2002. *Abstracts of Western Classics on Sexology* 西方性学名著提要. Jiangxi People's Press 江西人民出版社.

Li, Yinhe 李银河, ed. 2007. *Qixi, Folk Customs, and the Culture of Emotions* 七夕 / 民俗 / 情感文化. China Radio/TV Press 中国广播电视出版社.

Li, Yinhe 李银河. 1991. *Sexuality and Marriage of the Chinese* 中国人的性与婚姻. Henan People's Press 河南人民出版社.

Li, Yinhe 李银河. 1992–1993. *Their World – A Look into Chinese Male Homosexual Community* 他的世界 – 中国男同性恋群落透视. Hong Kong Cosmos Press 香港天地公司, 1992; Shanxi People's Press 山西人民出版社, 1993.

Li, Yinhe 李银河. 1993–1994. *Procreation and Chinese Village Culture* 生育与中国村落文化. Oxford University Press, Hong Kong, 1993; Chinese Academy of Social Sciences 中国社会科学出版社, 1994.

Li, Yinhe 李银河. 1995. *Transformation of Marriage and Family in China* 中国婚姻家庭及其变迁. Heilongjiang People's Press 黑龙江人民出版社.

Li, Yinhe 李银河. 1996. *Sexuality and Love of Chinese Women* 中国女性的性与. Oxford University Press, Hong Kong.

Li, Yinhe 李银河. 1997. *The Rising Power of Chinese Women* 中国女性利的崛起. Chinese Social Science Press 中国社会科学出版社.

Li, Yinhe 李银河. 1998. *The Subculture of Sadomasochism* 虐恋亚文化. Today's China Press 今日中国出版社.

Li, Yinhe 李银河. 1998. *Chinese Women's Emotions and Sex* 中国女性的感情与性. Today's China Press 今日中国出版社.

Li, Yinhe 李银河. 1998. *The Subculture of Homosexuality* 同性恋亚文化. Today's China Press 今日中国出版社.

Li, Yinhe 李银河. 1999. *Sex/Marriage – East and West* 性 / 婚姻 — 东方与西方. Shaanxi Normal University Press 西师范大学出版社.

Li, Yinhe 李银河. 1999; 2003. *On Sexuality* 性的. Chinese Youth Press 中国青年出版社.

Li, Yinhe 李银河. 2000. *Enjoy Life* 享受人生. Baihuazhou Arts Press 百花洲文艺出版社.

Li, Yinhe 李银河. 2000. *Queer Theory – the Ideological Trend in Sexology in the 1990s West* 酷儿理 – 西方90年代性思潮. Shishi Press 时事出版社.

Li, Yinhe 李银河. 2001. *Foucault and Sexuality – Reading Foucault's "The History of Sexuality"* 福柯与性 – 解福柯'性史'. Shandong People's Press 山东人民出版社.

Li, Yinhe 李银河. 2003. *Female Emotions and Sexuality* 女性的感情与性. Wunan Press 五南出版公司, Taiwan.

Li, Yinhe 李银河. 2003. *On Feminism* 女性主义. Wunan Press 五南出版公司. Taiwan.

Li, Yinhe 李银河. 2003. *Report on the Study of Sexual Culture* 性文化研究报告. Jiangsu People's Press 江苏人民出版社.

Li, Yinhe 李银河. 2003. *Sexual Love and Marriage* 性与婚姻. Wunan Press 五南出版公司, Taiwan.

Li, Yinhe 李银河. 2004. *Forum on Feminism* 女性主义坛. Proceedings of the Forum on Feminism 会文集.

Li, Yinhe 李银河. 2004. *The Poor and the Rich – the Diversification of Chinese Urban Families* 人与富人——中国城市家庭的富分化. Huadong Normal Uinversity Press 华东师范大学出版社.

Li, Yinhe 李银河. 2004; 2006. *Loving You is Like Loving Life* 你就像生命. Zhaohua Press 朝华出版社, 2004; republished under a different title: *Fall in Love if You Wish* 假如你愿意你就恋吧, Shanxi Normal University 西师范大学, 2006.

Li, Yinhe 李银河. 2004-2005. *On Gender Relations* 两性系. Wunan Press 五南出版公司, Taiwan, 2004.

Li, Yinhe 李银河. 2005. *On Feminism* 女性主义. Shandong People's Press 山东人民出版社.

Li, Yinhe 李银河. 2005. *You Are So In Need of Comfort – A Dialogue on Love* 你如此需要安慰 —— 于的对. Contemporary World Press 当代世界出版社.

Li, Yinhe 李银河. 2006. *Li Yinhe's Self-Selected Works* 李银河自选集. Inner Mongolia University Press 内蒙古大学出版社.

Li, Yinhe 李银河. 2007. *Live with Grace and Gentleness* 以温柔优雅的度生活. Chinese Women's Press 中国女出版社.

Li, Yinhe 李银河. 2007. *On Gender* 性别. Qingdao Press 青出版社.

Li, Yinhe 李银河. 2008. *20 Lectures on Sexual Love* 性20. Tianjin People's Press 天津人民出版社.

Li, Yinhe 李银河. 2008. *Li Yinhe's Reflections on Sexual Studies* 李银河性学心得. Times Arts Press 时代文艺出版社.

Li, Yinhe 李银河. 2009. *The Women of Houcun – Gender and Power in the Countryside* 后村的女人 – 农村性别利系. Inner Mongolia Press.

Li, Yinhe 李银河. 2009. *The Essence of Sociology* 社会学精要. Inner Mongolia Press 内蒙古出版社.

Li, Yinhe 李银河. 2014. *A Study of the Language of Sexuality in the New China* 新中国性研究. Shanghai Social Science Press 上海社会科学出版社.

Li, Yinhe 李银河. 2014. *Introduction to Sexology* 性学入门. Shanghai Social Sciences Press 上海社会科学出版社.

Li, Yinhe 李银河. 2014. *My Philosophy of Life* 我的生命哲学. Chinese Labor and Commerce Union Press 中华工商联合出版社.

Li, Yinhe 李银河. 2014. *My Social Observation* 我的社会察. Chinese Labor and Commerce Union Press 中华工商联合出版社.

Li, Yinhe 李银河. 2014. *Readings of My Soul* 我的心灵. Chinese Labor and Commerce Union Press 中华工商联合出版社.

Li, Yinhe. Friday sexology salon on TV. Accessed October 2015. http://z.t.qq.com/sexology/salon04.htm.

Li, Yinhe. *Li Yinhe Blog*. Accessed February 22, 2016. http://blog.sina.com.cn/liyinhe.

Liang, Shuming. 2006. *Eastern and Western Cultures and Their Philosophies* 东西文化及其哲学. First edition, 1922; reprint, Shanghai People's Press 上海人民出版社.

Liang, Zhongtang 梁中堂. "Arduous Journey: From One Child to Homes of Girls" 的历程：从 "一胎化" 到 "女儿户." *Open Times*

放时代 (online magazine). Accessed February 2016. http://www.opentimes.cn/bencandy.php?fid=375&aid=1806.

Liao, Xiaoyi. 2010. *Turning East and Looking West: Environmental Remedies: Sheri Liao's Talks with Eastern and Western Thinkers*东西望. SunChime.

Linshi, Jack. "Why This Chinese Startup Just Bought a Company Americans Love to Ridicule." *Time*. April 15, 2015. Accessed February 2016. http://time.com/3822962/segway-ninebot-china/.

Liu, Haiyan. "Hooligan Yan's Blog 流氓燕的博客." Fenghuang Blog. Accessed March 28, 2016. http://blog.ifeng.com/1403777.html.

Liu, Zhengzheng. "General Aviation, Next Trillion Dollar Market 通用航空，下一个万一市?" *CARNOC.com*. 民航源网. September 29, 2013. Accessed February 2016. http://news.carnoc.com/list/262/262568.html.

London Chinese. Accessed March 2016. http://www.londonchinese.com/forum.php?mod=viewthread&tid=17904.

Long, Pingping 龙平平, Zhang, Shu曙. "Deng Xiaoping Resolved to Reinstate College Entrance Examination, Changing the Fates of a Generation of Educated Youth 邓小平决策恢复高考，改变一代知青年命运." *News of the Communist Part of China*. Accessed February 2016. http://cpc.people.com.cn/GB/64162/64172/85037/85039/6032327.html.

McDougall, Bonnie S. Editor and Translator, *Notes From the City of the Sun: Poems by Bei Dao*. Ithaca, New York: China-Japan Program, Cornell University.

Ministry of Industry and Information Technology of the People's Republic of China Website 中华人民共和国工业和信息化部网页. Accessed February 2016. http://www.miit.gov.cn/n11293472/n11293832/n11294042/n12876231/16048026.html.

Murong, Sujuan 慕容素娟. "*Integrated Circuit Outline* Reading Chen Datong: Government Foundation Allied with Private Capital, Holding up the Sky for the IC Industry. 《集成电路要》解之大同：政府基金与民间本联手 中国 IC工业擎起一片天." Chinese Electronic Information Enterprise Web 中国电子信息业网, *China Electronic Paper* 中国电子报. Accessed February 2016. http://semi.cena.com.cn/2014-07/04/content_231514.htm.

Naughton, Barry. 2007. *The Chinese Economy: Transitions and Growth*. Cambridge: MIT Press.

Nayler, Mark. "Ready for Take-off." *Spear's Asia*, No. 3, 2012/13. Accessed February 2016. http://www.asianskygroup.com/attachment/news/25/SPEARS-asia-ASG.pdf.

Newport, Frank. "Six out of 10 Americans Say Homosexual Relations Should Be Recognized as Legal." *Gallup News Service*. May 15, 2003. Accessed March 2016. http://www.gallup.com/poll/8413/six-americans-say-homosexual-relations-should-recognized-legal.aspx.

Office of Educational Affairs of the Embassy of the P.R. China in the US, "General Introduction to Overseas Study 出国留学介." Accessed February 2016. http://www.sino-education.org/policy/studybrief.htm.

OmniVision web page. Accessed February 2016. http://www.ovt.com/technologies/.

Osnos, Evan. 2014. *Age of Ambition: Chasing Fortune, Truth, and Faith*. New York: Farrar, Straus and Giroux.

Pan, Jiutang 潘九堂, Hui, Liu 刘, and Quan, Yuan 袁泉. "An Analysis of the Origins and the Competitiveness of Shenzhen's Shanzhai Cell phone Production System 深圳山寨手机生体系的起源和争力分析." Accessed November, 2015. http://www.ide.go.jp/English/Publish/Download/Jrp/pdf/156_ch1.pdf.

Peekabook Chinese-English Children's Library 皮卡屋中英文少儿. Blog website 新浪博客. Sina.

Petti, Claudio, ed. 2012. *Technological Entrepreneurship in China: How Does It Work?* Edward Elgar Publishing Limited.

PRC State Department, Ministry of National Defense of the People's Republic of China. 2014. "Maintaining Airspace Security is a Common Responsibility of the Entire Society 护空中安全是全社会的共同任." *PRC Ministry of National Defense Website*. Accessed February 2016. http://news.mod.gov.cn/pla/2014-02/25/content_4492494.htm.

Qing, Jing. "Statistics of Annual Exit and Return Students (2008–2013) 中国历年出国回国留学人数统 (2008–2013)." *360doc*. March 2, 2014. Accessed February 2016. http://www.360doc.com/content/14/0302/21/5177773_357178235.shtml.

Qinghua University. 2004. *On Commercial Law* 经济法概. Qinghua University.

Qu, Kai 瞿. "60 Years of CAAC: A Brief Introduction to the Development of China's General Aviation" 民航60年：中国通用航空

发展概况." *CAAC Resource Web* 民航源网. Accessed October 2015. http://news.carnoc.com/list/145/145212.html.

Seki, Mitsuhiro. 1994. *Beyond the Full-Set Industrial Structure: Japanese Industry in the New Age of East Asia.* Tokyo, Japan: LTCB International Library foundation.

Spence, Jonathan. 1990. *The Search for Modern China.* New York: Norton.

State Department, Central Military Committee, People's Republic of China. 2010. "Opinions On Deepening the Reform on the Management of Our National Low-altitude Airspace 于深化我国低空空域管理改革的意见." *Economic Observation Web* 经济察网. Accessed February 2016. http://www.eeo.com.cn/industry/shipping/2010/11/16/185929.shtml.

State Department, Central Military Committee, People's Republic of China. "The General Flight Rules of the People's Republic of China 中华人民共和国飞行基本." *Beijing University Legal Information Website.* Accessed October 2015. http://www.lawinfochina.com/display.aspx?lib=law&id=11871&CGid.

State Department, Central Military Committee, People's Republic of China. 2003. "Regulations on the Flight Management of General Aviation 通用航空飞行管制条例." *Office of PRC State Department, Government Public Announcement Website.* Accessed February 2016. http://news.carnoc.com/list/266/266869.html.

State Department, Central Military Committee, People's Republic of China. 2003. "Regulations on the Flight Management of General Aviation 通用航空飞行管制条例 (English translation)." *Beijing University Legal Information Website.* Accessed February 2016. http://www.lawinfochina.com/display.aspx?lib=law&id=2613&CGid.

State Department, Central Military Committee, People's Republic of China. "Regulations on Approval and Management of General Aviation Flight Plans 通用航空飞行任务批与管理定." *CAAC Resource Web* 民航源网. Accessed February 2016. http://www.caac.gov.cn/H1/H3/. 2013.

State Department, People's Republic of China. "Temporary Regulations on the Management of the General Aviation 国务院于通用航空管理的时定." (Online) *Law Library* 法律. January 8, 1986. Accessed February 2016. http://www.law-lib.com/law/law_view.asp?id=3484.

Steinfeld, Edward. 2010. *Playing Our Game*. Oxford.

Sun Chang, Kang-I and Owen, Stephen, eds. 2010. *The Cambridge History of Chinese Literature Vol. 2: From 1375*. Cambridge University Press.

Wang, Huiyao. 2005. *Returnee Time* 海归时代. Central Translation Press.

Wang, Huiyao. 2007. *Contemporary Chinese Returnees* 当代中国海归. Beijing: Zhongguo Fazhan Chubanshe.

Wang, Huiyao. 2012. *Blue Book of Global Talent: A Report on the Development of Chinese Studying Abroad (2012) No. 1*. 国际人才皮：中国留学发展报告. Social Science Archive Press 社会科学文献出版社.

Wang, Huiyao. 2012. *Globalizing China: The Influence, Strategies and Successes of Chinese Returnee Entrepreneurs*. Emerald Group Publishing Ltd.

Wang, Yingchun. "Lehe Home: An Idealist's Village Building Utopia." *Journal of Chinese Economy*, October 27, 2010.

Wikiquote. "... it is cheap to go for a $100 prostitute ..." Accessed October 2015. http://en.wikiquote.org/wiki/Han_Han.

Wong, Edward. "Beijing Issues Red Alert Over Air Pollution for the First Time." *New York Times*. December 7, 2015. Accessed February 2016. http://www.nytimes.com/2015/12/08/world/asia/beijing-pollution-red-alert.html.

World Affairs Council: Oregon. "This Endangered Planet: A Chinese View." *FOR A.tv*. Accessed March 29, 2016. http://library.fora.tv/2007/05/29/This_Endangered_Planet_A_Chinese_View.

Xiang, Zhaoqin 向兆琴. "Research Found Sales of Chinese Shanzhai Cell Phones Reached 145 Million Sets 查示09年中国山寨手机售量达 1.45亿部." *Tencent*. November 6, 2009. Accessed November 2015. http://tech.qq.com/a/20091106/000228.htm.

Xie, Lei, ed. "Understanding Shanzhai: Is It Grassroots Spirit or a Killer of Innovation? 解"山寨"：草根精神抑或新"手"？" *News of the Communist Party of China*. December 31, 2008. Accessed February 2016. http://theory.people.com.cn/GB/40555/8610266.html.

Xinjingbao 新京报. "Environmental Protection Minister: Has Seen 'Under the Dome,' Texted to Thank Chai Jing 环保部部长：已看过《穹之下》，短信感了柴静." *Sohu*. March 1, 2015. Accessed February 2016. http://news.sohu.com/20150301/n409242625.shtml.

Xu, Huitan 徐会坛. "Liao Xiaoyi: From a Western-Styled
Environmentalist to a Oriental-style Environmentalist 从西方环保
者到东方环保者." *iCixun*. March 5, 2015. Accessed February 2016.
http://www.icixun.com/2015/0305/4855.html.

Ye, Xiaonan 叶晓楠. "Beijing: the Plan for Capturing Talents Under
the Government's Directive 北京：政府主下的才划." *People's Daily*
(Overseas edition) January 10, 2009. People Net. Accessed February
2016. http://paper.people.com.cn/rmrbhwb/html/2009-01/10/
content_173333.htm.

Yoon, Kwangho, Kim, Chanki, Li, Bumha, and Lee, Doyong.
"Single-Chip CMOS Image Sensor for Mobile Applications." *IEEE
Journal of Solid-State Circuits* 37:12. Abstract at: *IEEE Xplore Digital
Library*. Accessed February 2016. http://ieeexplore.ieee.org/xpl/login.
jsp?tp=&arnumber=1088114&url=http%3A%2F%2Fieeexplore.
ieee.org%2Fiel5%2F4%2F23646%2F01088114.
pdf%3Farnumber%3D1088114.

You, Tracy. "Flying Dragons: Private Jets Are New Status
Symbol in China." *CNN Travel*. April 11, 2012. Accessed
October 2015. http://travel.cnn.com/shanghai/life/
chinas-business-jets-reach-new-heights-113799.

Zhang, Lei. "Understanding the Intellectual Property Rights of the
'Shanzhai' phenomenon '山寨' 象的知法解." 2011. *Wanfang Data*.
Accessed February 2016. http://d.wanfangdata.com.cn/Thesis/
Y2049876.

Zhao, Suisheng, ed. 2006. *Debating Political Reform in China: Rule of
Law vs. Democratization*. M. E. Sharpe, Inc.

Zhong, Ming, ed. "The Capital City Blossoms with Private Children's
Libraries, Games Cultivate Reading Habit 京城兴起私人儿童，游
戏培." *Cultural China*. August 23, 2010. Accessed March 28, 2016.
http://culture.china.com.cn/book/2010-08/23/content_20766338.
htm.

Index

FAA (Federal Aviation Administration), 95
family, 59, 61, 65, 69, 121, 123, 133, 157, 187, 191
feminist, 59
Foucault, Michel, 59, 71
founder, 21, 117, 119, 125, 135, 141, 151, 165, 191
Four Modernizations, 19, 189
funding, 41, 49, 69, 149, 151, 159, 183

GA (General Aviation), 19, 87, 89, 91
gay rights, 23, 57, 59
GDP, 179, 183
gender equality, 177
generation, 9, 19, 25, 57, 61, 71, 75, 135, 159, 161, 171, 193
global challenge, 147
global citizen, 49
global economy, 17, 23
GMD (Guomindang), 167
government regulation, 65
Gu Cheng, 11
GVB (Global Village of Beijing), 189

harmony, 21, 51, 189
health, 25, 59, 93, 191
helicopter, 89, 93, 97, 103
heterosexual, 61
hierarchy, 126, 172
hi-tech, 21, 141, 179, 181, 183
Hong Kong, 9, 95, 97, 99
hooliganism, 61, 63
Hu Birong, 117, 119, 191
Hua Capital, 179, 181
human rights, 81
Hurun List, 111

IC (integrated circuit), 147
incubator, 141
individualism, 77, 93, 99, 131, 135, 171, 175, 193
industrialization, 23, 35, 51

innovation, 159, 165
Internet, 177
invention, 169
investor, 164n6, 166n13, 166n16, 188
IPO (Initial Public Offering), 151, 185

jet owners, 184

Keck, Elizabeth, 95
King's Aviation, 89, 91, 107

law, 17, 23, 59, 61, 63, 65, 67, 71, 77, 177
leadership, 49, 61, 143, 145, 185
legalization, 17, 23, 73
Lehe, 37, 39, 43, 45, 49, 189
LGBT, 57, 59, 69
Li Yinhe, 17, 23, 55, 57, 61, 67, 175, 177
Liang Suming, 37
Liao Xiaoyi, 31, 43
library, 21, 25, 117, 119, 121, 123, 125, 127, 129, 133
love, 47, 61, 67, 69, 71, 131, 133, 175, 177
Luo Ming, 117, 133
Luo Qun (Meggie), 155

management, 39, 49, 99, 147, 151, 179
Mao Zedong, 135
market economy, 71–74
marriage rights, 175
May Fourth Movement, 61, 69
media, 23, 25, 27, 35, 39, 75, 77, 119, 125,
merger, 180
meritocracy, 13
migrant workers, 47, 49, 135, 183
military, 87, 95, 99, 107, 113
modernization, 17, 19, 21, 27, 63, 183, 189,

national entrance exam, 9
national security, 87, 181

photo © Rebecca Andersen

ABOUT THE AUTHOR

Zhuqing Li, PhD, is adjunct associate professor at Brown University. Her research and teaching range from linguistics to literature and culture in East Asia. Her major academic publications include three books on Chinese linguistics. She has taught courses on East Asian culture in major universities for fifteen years. This is the first book-length study of a group of foreign-educated elite instrumental in China's rise in the course of the past thirty years.